Bursting Neurons and Fading Memories
An Alternative Hypothesis of the Pathogenesis of Alzheimer's Disease

Bursting Neurons and Fading Memories
An Alternative Hypothesis of the Pathogenesis of Alzheimer's Disease

Michael R. D'Andrea

AMSTERDAM • BOSTON • HEIDELBERG • LONDON
NEW YORK • OXFORD • PARIS • SAN DIEGO
SAN FRANCISCO • SINGAPORE • SYDNEY • TOKYO

Academic Press is an imprint of Elsevier

Academic Press is an imprint of Elsevier
32 Jamestown Road, London NW1 7BY, UK
525 B Street, Suite 1800, San Diego, CA 92101-4495, USA
225 Wyman Street, Waltham, MA 02451, USA
The Boulevard, Langford Lane, Kidlington, Oxford OX5 1GB, UK

Notices
Knowledge and best practice in this field are constantly changing. As new research and experience broaden our understanding, changes in research methods, professional practices, or medical treatment may become necessary.

Practitioners and researchers must always rely on their own experience and knowledge in evaluating and using any information, methods, compounds, or experiments described herein. In using such information or methods they should be mindful of their own safety and the safety of others, including parties for whom they have a professional responsibility.

To the fullest extent of the law, neither the Publisher nor the authors, contributors, or editors, assume any liability for any injury and/or damage to persons or property as a matter of products liability, negligence or otherwise, or from any use or operation of any methods, products, instructions, or ideas contained in the material herein.

British Library Cataloguing-in-Publication Data
A catalogue record for this book is available from the British Library.

Library of Congress Cataloging-in-Publication Data
A catalog record for this book is available from the Library of Congress.

ISBN: 978-0-12-801979-5

For information on all Academic Press publications
visit our website at http://store.elsevier.com/

This book has been manufactured using Print On Demand technology.

www.elsevier.com • www.bookaid.org

About the cover: The exploding star (left) and exploding neuron (right) produce the supernova Cassiopeia A and the dense-core amyloid AD plaque, respectively. Both images show the residual debris: the remnants of the star (cyan) and neuronal nucleus (purple) are located at the same 10:00 position in their respective photographs. Supernova photograph is courtesy of NASA/JPL-Caltech/ O. Krause (Steward Observatory). AD plaque figure is courtesy of Current Pharmaceutical Design, 2006;12(6):677–684.

DEDICATION

I dedicate this book to:

- Those who passed away from this dreadful disease and those who caringly donated their loved ones' tissues to research. I would not have discovered so much without their ultimate contributions. Those who have AD and their family's caregivers, for help is soon on the way.
- My wife Patty, my oldest daughter Dr. Michelle and her husband, Kevin, my son Michael, and my youngest daughter, Stephanie, all of whom have given me endless support and love.
- My parents, Henry and Angela, for all their eternal love and support, and especially for buying my first microscope as a Christmas gift when I was 7 years old, setting my scientific career in motion.

TABLE OF CONTENTS

Michael R. D'Andrea received his PhD in cell and developmental biology, his MS in molecular biology at Rutgers University, New Brunswick, NJ, and his BA in psychobiology at Western Maryland College, Westminster, MD. His dissertation work utilized molecular and histological assays to study the regulation of DNA topoisomerases in human cancers. His earlier career concerned the use of the high-magnification electron microscope to support oncogenesis in preclinical models, and then moved into a new field where he and his peers coinvented the chorionic villus sampling method for clinical chromosomal analysis at Thomas Jefferson Medical School. In the late 1980s and early 1990s, he mastered immunohistochemical methods at the light and electron microscopy levels when he began publishing his work in scientific journals. However, it wasn't until the mid-1990s, while working at Johnson & Johnson's Pharmaceutical Research & Development as the Target Validation Team Leader, that he became engaged in Alzheimer's disease (AD) research. The team that he established was responsible for supporting target discovery and validation, while supporting biomarker discovery in preclinical and experimental models using genomic, proteomic, and histopathological methods across many therapeutic areas. For this work, he was honored with over a dozen Leadership and Scientific awards.

Currently, he has over 100 peer-reviewed scientific publications and invited reviews, about one-third of which concern the neuropathology of AD, and holds 11 scientific patents. He has reviewed hundreds of papers and served on several editorial boards for many scientific journals, has reviewed international grants in the AD field, and is currently on the editorial board of the journal *Biotechnic & Histochemistry*. He has been invited to speak at numerous international, national, and regional meetings, as well as at universities and other companies to discuss his novel observations concerning the origin of amyloid plaques, the existence of various plaque types, and most recently the assertion that AD is also an autoimmune disease, all of which are presented in this book. Most recently, Michael established a contract research company, Slidomics, LLC (www.slidomics.com), to apply his histopathological and target validation expertise by providing high-quality data and analysis much like what you see in this book.

PREFACE

So how do you write a book about Alzheimer's disease (AD)? You can state the facts, or analyze the progress the field has made so far, or discuss your personal experiences. But how do you write a book about a possible cure for AD? Quickly, I suppose. The sooner progress can be made, the better.

When I first had the idea for this book, I was shocked that it took so long for the thought to come to me. By that time I had given up on having any impact in Alzheimer's research, despite my over 30 publications, many presentations, and 16 years in the field. Since my research has contradicted the mainstream hypotheses in the field, hopefully those of you who do know what that is like can sympathize with my distress and frustration at the current state of affairs. But to finally write a summary of these findings in one cohesive book now seems obvious, and writing it came so naturally that I only wish I embarked on this journey sooner, if only because it will never be too soon to find a cure for AD.

While I have very high hopes for how this book might change the direction of AD research, at the very least it has already been a huge personal relief to write, if only to say "look back over here, I think I've got it." It's a bit dramatic perhaps to call this my "swan song," but I think of it more as passing on the torch, hopefully to a generation of bright, passionate, unbiased, but skeptical scientists who can continue my legacy of research. The feeling that years of personal emotional investment into this field might still be valuable is my relief.

And if you're in the field, you must know that now, more than ever, there is an imperative need to consider alternative hypotheses about the causes of AD. The field needs controversy to move forward and I certainly supplied my fair share since the late 1990s. If I wasn't completely confident in these personal findings and the direction of my hypothesis, I'd remain quiet and frustrated as you may be with the failed "all-eggs-in-one-basket" approach that has stifled creative thinking in the field to explain neuronal death leading to a cure. I hope reading about this hypothesis not only exposes the flaws in the current line of research but also convinces you exactly where to go next. And if you don't believe what I've found, please conduct research to support or contradict me! Again, this field needs new ideas, no matter where they come from or what they are, and now that the oppressive cloud of the reigning hypothesis is dissipating, it is time to consider alternative hypotheses.

Writing this book and recalling the story of my many contributions to the field has been an emotional experience. I have relived the pride in designing one set of assays after the next, the excitement of reading the hundreds of microscopic slides of human tissues, and the frustration of the political challenges I faced by publishing controversial data. As you will discover, I never had a passion to cure AD when the story began, nor did I have any stakes in the field as a cell biologist. I was merely the lucky scientist who came across the provocative results I present here. I regret that I could not make a stronger impact previously through each of my publications and presentations, but I hope I can now by collating this work into one coherent and logical story. I encourage you to read with a critical mind, dig up these detailed publications, question everything you know or have learned about AD, and consider what must happen for the field to move forward and what you can do to help. Although I have addressed the many editorial comments of the reviewers and audience members I have faced, as you will read, I will depend on you to assess the scientific value of my work on its own merits, in the context of the current state of the AD field.

Lastly, thank you for sharing in my journey. It is not easy being the odd one out, and here more than ever I am putting myself up for criticism and disdain. Please know that everything I present is for the betterment of the field, for our futures, and hopefully serves to move us more quickly toward a cure for this tragic, draining, and terrifying condition known as AD.

Conformity may give you a quiet life; it may even bring you to a University Chair. But all change in history, all advancement, comes from the nonconformists. If there had been no trouble-makers, no Dissenters, we should still be living in caves.

AJP Taylor

Disclosure statement: Although all of the data presented in this book have been published in peer-reviewed scientific journals, I make no claims on how to treat AD patients beyond what is already known in the literature. The sole intention of this book is to present novel hypotheses for consideration in light of the failed clinical studies based on the classical amyloid cascade hypothesis. As with any scientific trials, if the hypothesis is proven false, then other hypotheses should be considered for testing.

ACKNOWLEDGMENTS

Especially to my immediate family for providing direction, style, and edits for the book.

Dr. Bruce Damiano, Dr. Charlie Saller, and Peter Cronk, Esq., for I am also indebted to them for their keen edits, constructive criticisms, and support to move ahead with this project.

To all of my past collaborators, hopefully they know who they are and appreciate the contributions they have made to my line of research and the AD field.

To the Elsevier publishing staff, especially Dr. Natalie Farra for her supportive contributions to this book.

Thank You!

INTRODUCTION

... dead ends in Alzheimer's research ...

How has Alzheimer's disease (AD) affected your life? Is it in your profession to directly care for or treat AD patients? Are you in the scientific and medical community trying to discover the cause? Do you have a loved one who is currently suffering from this terrible disease? Or, perhaps, you have a sideline curiosity or simple fascination with AD. Even if AD does not impact your life today, the odds are sadly staggering that it will someday.

AD diagnoses are increasing at an alarming rate: today, as many as half of people over 80 will be afflicted.[1] AD is officially the sixth leading cause of death in the United States and fifth leading cause of death for those of ages 65 and older; that is more than prostate cancer and breast cancer combined.[2] In other words, the odds are high that your parents, siblings, other relatives, and/or neighbors will be diagnosed with AD as they age. To have a loved one not only forget you but also require full-time care over the course of several, perhaps many years can cripple any human spirit, as some of you undoubtedly and unfortunately already know.

It is impossible to overstate the urgency and dire state of AD today, and there are no signs of slowing down. Deaths from AD increased 68% between 2000 and 2010, and AD is among the top 10 causes of death in America that cannot be prevented, cured, or even slowed.[2] About 13.8 million Americans will be living with AD by 2050, up from 4.7 million in 2010, and according to the World Health Organization, about 35.6 million people around the world have dementia, with 7.7 million new cases each year.[3] Imagine that every 67 seconds, someone develops AD.[2] This disease not only negatively impacts the immediate family and friends of the victim but also is one of the most costly modern medical conditions to support. In 2014, the direct costs to American society for AD care will total an estimated $214 billion, and if there is no breakthrough cure or way to prevent or even slow down the progression of AD, the costs may reach up to a staggering $1.2 trillion by the year 2050.[2]

If there is still no cure for or better understanding of AD today, it's not for lack of trying. In the United States alone, government initiatives have funded $2.5 billion in AD research just over the past 4 years including a projected $566 million in 2015.[4] Despite these impressive numbers, researchers dedicated to the field have had few breakthroughs. Most recently, a series of high-profile, expensive clinical trials hedging their bets on one hypothesis have all failed, leading to a dead-end in the field. These recent failures have fueled a mounting frustration and a lack of confidence in the current direction of the study of AD. In the face of a stagnant field, with all prominent and heavily funded current initiatives in the field being uncompromising failures, this book is intended to be the impetus for change, for a new approach, and for hope to combat this terrible condition. With this book, I hope to change your perception of the causes of AD, and that you will come to understand how I developed this novel hypothesis, which will logically explain why research has reached a dead-end while presenting a road map for alternative approaches.

All of the evidence supporting this hypothesis concerning the causes of AD is presented in a general chronological order (not a scientific or "textbook" order) with actual events. Each study is accompanied with rationale, commentary, technical and observational details, and interpretations of these experiments. The spirit of the book is not a data download filled with hundreds of references, but a story of my journey of how I was forced to confront peers in a very political field with data contradicting their long and dearly held beliefs. Again, this is a story, not a journal article, and I do not mean for this book to be the end-all AD reference book. Although there are many more thorough sources on the history of AD than I can possibly cover in this book, I can summarize some of the important aspects of AD to help tell my tale.

I sincerely hope that as you read this book, it gives you a new perspective on the causes of neuronal death leading to AD, and that if you are in the field, it stimulates you to pursue and test some of these findings to help cure this disease. I hope that you find my stories invigorating, informative, provocative, and yet frustrating in the face of a politically charged academic system primarily based on old science. I encourage you to approach this book with an open mind and little bias, and only ask that you refrain from enacting your commentary until you complete the reading. Please remember that the scope of this edition is narrow, and that it is impossible to incorporate years of related literature into this book. Although the presentation is narrow, the focus is clear. I will first take you through the background of AD, help you understand the current state of the field, and then take you through a story of discovery, enlightenment, and the years of struggle that finally brought me to this book.

Concerning the figures throughout the book, I sincerely hope you appreciate all of the photomicrographs. They are the true treasures of this book that form the basis of this unique hypothesis, and, most importantly, they represent those who donated their loved ones' tissues for scientific research. One other note about the images, some of which were featured on journal covers: please see those specific journal articles for the original figures to see them in their true scientific context, accompanied with scale bars, magnifications, methodological details, and plenty of relevant references. I am thankful for those specific journals, NASA, and Optos® to allow me permission to exhibit them in this book.

And with that, let's begin ….

REFERENCES

1. http://www.everydayhealth.com/alzheimers/alzheimers-risk-factors.aspx (accessed July 24, 2014).

2. www.alz.org (accessed June 1, 2014).

3. http://www.who.int/mediacentre/news/releases/2012/dementia_20120411/en/ (accessed August 2, 2014).

4. http://report.nih.gov/categorical_spending.aspx (accessed June 30, 2014).

CHAPTER 1

Alzheimer's Disease Today

... dashing hopes for millions ...

Alzheimer's disease (AD), or Alzheimer disease, is a progressive, neuro-degenerative disorder of the brain that mysteriously claims the lives of millions of people. It leaves the affected helpless, the community frightened, and the AD research field frustrated.

Mental deterioration in old age has been recognized and described throughout history. However, it was in 1906 that Dr. Alois Alzheimer defined the severe aspects of this condition, based on the autopsy of one of his patients who had died after years of severe memory problems, confusion, and difficulty understanding questions. He described the presence of dense deposits surrounding the nerve cells and twisted fibers in the nerve cells in the brain. Today, we refer to these dense deposits as plaques, areas of complex proteins that deposit in the AD brain. Those plaques found in the AD brain are generally composed of amyloid, an insoluble, fibrous protein. The twisted fibers that Dr. Alzheimer described, now known as neurofibrillary tangles, are abnormally constructed tau proteins, otherwise meant to stabilize the cell cytoskeleton for normal cellular activities. Together, the presence of these neuropathological features confirms the diagnosis of AD; however, it is not agreed upon if these are the cause of neuronal death leading to AD or a by-product of it.

Clinically, the manifestations of AD are also fairly well defined. AD begins with symptoms of dementia, although not everyone with dementia regresses to AD. These symptoms include difficulty with language, memory, perception, thinking, and judgment, and can deteriorate into a stage termed mild cognitive impairment (MCI). The additional symptoms of MCI include difficulty in solving problems, multitasking, and forgetting recent events or conversations, although, again, not everyone with MCI will develop AD. The early stages of AD occur if these symptoms continue to worsen to the point of getting lost on familiar routes, having language problems, losing interest in previously enjoyed activities, and, finally and most tragically, personality changes and the loss

Bursting Neurons and Fading Memories. http://dx.doi.org/10.1016/B978-0-12-801979-5.00001-1

of social skills. As the AD progresses, the symptoms worsen to include hallucinations, unawareness of self, and the inability to be independent, thereby requiring full-time care. This cruel disease is equally troubling to family and friends who agonize over the vacant expression on their loved one's face and are left to painfully watch life slip away. AD finally claims the victim's life about 8–10 years after the initial diagnosis, chiefly due to coexisting illnesses such as pneumonia.[1]

The cause of the death of the neurons in the brain is unknown but several recent discoveries have brought to light additional information regarding the pathological basis for AD. Some of these form the basis for hypotheses on AD, the most prominent and influential of which will be illustrated later.

I will take this opportunity to go into more technical detail concerning the mechanisms of the brain and neurons. I cannot fully describe or explain these nuances here, but if you are not familiar with all the terminology used, rest easy knowing that fully understanding this section is not necessary to appreciate the rest of the text. Also, if you are unfamiliar with some terms or would simply like a refresher, please find a brief technical glossary at the end of this book.

As an additional disclaimer, I don't intend to present myself as an expert on any of the preceding proposed AD mechanisms. While the information I present is accurate, I apologize in advance to specialists in the field for any egregious oversimplifications or omissions made for the sake of conciseness.

NEUROLOGICAL FACTORS

While it has been known since Dr. Alzheimer's finding in 1906 that amyloid plaques and abnormally constructed tau protein fibers are neuropathological features in the AD brain, only relatively recently has the scientific community begun to learn more about additional factors underlying neuronal death in the disease.

One of the first major discoveries in the 1970s was that of a deficit in choline acetyltransferase, an enzyme that synthesizes the neural transmitter acetylcholine, in the neocortex of the AD brain. Additional studies reported reduced choline uptake, increased acetylcholine release, and the degeneration of cholinergic neurons (those that use acetylcholine as a neurotransmitter) in the specific areas in the AD brain.[2-7] These data make up the foundation of the cholinergic hypothesis that the loss of cholinergic neurons, and thus the loss of cholinergic neurotransmission in critical brain areas, contributes significantly to the deterioration in cognitive function of AD patients.[8] The contemporary discovery of acetylcholine's pivotal role in learning and memory further supported this hypothesis.[9] Today, the cholinergic hypothesis is the basis of most of the currently available drug therapies to treat AD, which are meant to inhibit cholinesterase, another enzyme that breaks down acetylcholine. As of today, these therapies have had minimal success.[10]

Another prominent discovery in the 1980s involved neuroinflammation as the cause of neuronal death in the AD brain. In fact, the discovery of a wide array of immune-related antigens in the AD brain helped establish the concept of a specialized immunodefense system in the central nervous system. In particular, as a result of some factors in the AD brain (these are not universally agreed upon), microglia are reactive, setting off a chain of events releasing immune-related antigens including proinflammatory cytokines and chemokines.[11] According to the inflammation hypothesis, the increased secretion of these substances, which are potentially neurotoxic, eventually destroys neurons, leading to the development of AD symptoms.[12] Some proponents of the inflammation hypothesis also suggest that this sequence triggers the distortion of tau via phosphorylation.[12] Overall, the role of inflammation in AD is still widely debated. However, clinical trials targeting inflammation have not been impressive.[13]

While it is clear that phosphorylated tau (Dr. Alzheimer's twisted fibers) is somehow involved in the pathogenesis of AD, as either a cause or effect, its exact role is still heavily debated. Many researchers in the field believe that genetic mutations that alter the function and isoform expression of tau, a stabilizing neuronal filament, may initiate a far-reaching transformation that sets off AD. In this model of the tau hypothesis, it is reported that uncontrolled excessive or abnormal hyperphosphorylation of tau results in the transformation of normal adult tau.[14] This distorted adult tau begins to pair with other threads of tau to eventually form paired helical filaments, creating neurofibrillary tangles inside neurons. This has disastrous consequences, beginning with the collapse of the internal neuronal transport system and ending with cell death.[15-17] It was suggested that this may be the first malfunction in biochemical communication between neurons and later in the death of the cells.[18] It was reported that such impaired microtubule assembly due to the hyperphosphorylation of tau was found in AD brain extracts.[19] Subsequent research over the past 20 years has studied tau phosphorylation and function, and how site-specific phosphorylation modulates the physiological and pathological roles of tau. However, targeting hyperphosphorylated tau is difficult since it is inside neurons. One potential treatment is to administer modified methylene blue, which seems to prevent tau from aggregating to form those tangles.[20] This treatment showed enough promise in one Phase 2 trial to proceed to Phase 3 clinical trials with a related derivative. Other related trials are underway for drugs that stop phosphorylation of tau and antibodies against tau.[21]

NON-NEUROLOGICAL FACTORS

Besides discoveries regarding the neuropathology of AD, the AD field has also uncovered non-neurological risk factors and associated features of AD.

Perhaps the most straightforward discovery is the connection of vascular risk factors to the development of AD. In fact, the vascular hypothesis suggests that the pathology of AD begins with hypoperfusion, a decreased blood flow to the brain. Support for a vascular cause of AD comes from epidemiological, neuroimaging, pathological, and clinical studies.[22] The hypothesis considers cerebral microvascular pathology and cerebral hypoperfusion as primary triggers for neuronal dysfunction leading to the cognitive and degenerative changes in AD.[23] Advancing age and the presence of vascular risk factors create a critically attained threshold of cerebral hypoperfusion that ultimately leads to capillary degeneration.[14] Thus, the pathological consequences of capillary degeneration result in the development of plaques, inflammatory responses, and synaptic damage leading to the manifestations of AD.[24]

Genetics have also been implicated in the development of AD. Those who have a parent or sibling with AD are more likely to develop the disease and this probability continues to increase if more than one relative have or had AD. Although this suggests that AD has a genetic component, these currently known genetic risks account for only 0.1% of AD cases. The most prominent genetic risk factor is the gene that codes for apolipoprotein E (APOE4).[25] The APOE2 and APOE3 gene forms are the most common in the general population. It is the APOE4 gene that is associated with an individual's risk for developing late-onset AD. These lipoproteins are responsible for packaging cholesterol and other fats, and for carrying them through the bloodstream. Apolipoprotein E (ApoE) is also a major component of a specific type of lipoprotein, so-called very-low-density lipoproteins, which remove excess cholesterol from the blood to the liver for processing. Maintaining normal levels of cholesterol is essential for the prevention of disorders of the cardiovascular system, including hypertension, heart attack, stroke, and hypercholesterolemia, all of which are AD risk factors.[26] ApoE also has a role in neuronal signaling, and the maintenance of the integrity of the blood–brain barrier (BBB) that regulates the entry of selective substances into the brain. However, the exact pathophysiological process is yet to be elucidated.

The role of cholesterol in the pathology of AD is also shown by the ability of statins (drugs made to lower cholesterol levels) to reduce the prevalence of AD by up to 70%. Intracellular cholesterol may regulate amyloid processing by directly modulating the activity of secretase, the enzyme that breaks down the amyloid protein (see more in the section "The Amyloid Cascade") into smaller parts. Cholesterol may also affect the intracellular trafficking of amyloid and secretase.[27] In particular, high intracellular cholesterol increases γ-secretase activity and amyloidogenic pathways, while low intracellular cholesterol favors nonamyloidogenic pathways. There are also genetic factors linking cholesterol metabolism and AD. Although APOE is the only gene with replicable evidence, several candidate genes involved in lipid metabolism are being investigated for putative roles with mixed results.[28] Similarly, inhibition of cholesterol biosynthesis by statins and another cholesterol synthesis inhibitor was found to reduce amyloid burden in guinea pigs and murine models of AD.[14] Administration of inhibitors of cholesterol synthesis may reduce the prevalence of AD, although no firm conclusions have been reported.[29] However, several prospective studies have demonstrated that statins reduce the turnover of brain cholesterol at standard therapeutic doses, although the steady-state levels of Aβ, the form of amyloid most associated with AD, in the cerebrospinal fluid (CSF) remain unaltered. It is clear to see that the role of statins in AD requires further investigation.[28] Overall, biochemical and pharmacological evidence supports a role for cholesterol and lipid metabolism in amyloid. Ultimately, a complete understanding of how cholesterol can influence AD pathogenesis is unclear even though the efficacy of the statins looks promising.[30,31]

THE AMYLOID CASCADE

While I have presented many research initiatives, some of the most prominent and pertinent in the field, I do not mean for my list to be exhaustive. Some of these may come up later in the context of my own research findings, but for now I hope they serve to paint a general picture of the AD field as it stands today.

Of course, as anyone involved in AD research can tell you, this picture is woefully incomplete without addressing the most influential and widely embraced hypothesis of AD pathology, the amyloid cascade hypothesis. Before describing this possibility in detail, I would like to take a moment to bring you up-to-date on the field's understanding of the now notorious amyloid protein.

Amyloid is a fascinating topic but not quite easy to explain, since the literature refers to a family of names that describe small portions of the amyloid protein, some of which are purportedly more toxic than others. In a broad sense, amyloid refers to a group of "largely unrelated, pathologic, insoluble, extracellular, fibrous proteins" that usually derive from precursors present in the blood.[32] It wasn't until the accidental discovery that a particular stain called Congo red can selectively stain amyloid that it became possible to associate it with several diseases such as amyloidosis, rheumatoid arthritis, and AD.

In 1990, a guideline for nomenclature and classification of amyloid and amyloidosis was created and includes over a dozen names based on their origins and clinical associations. Of these names, the one associated with AD is the β-amyloid, frequently named Abeta (Aβ). This form of amyloid is derived from the amyloid precursor protein (APP) that represents a large family of cell membrane proteins. The normal function of Aβ is also not well understood; some report its role in synapse and memory formation, while others report its association with memory loss and neuronal cell death. At normal physiological levels, Aβ is a normal, soluble product of neuronal metabolism that regulates synaptic neuronal survival.[33] However, with age and for unknown reasons, Aβ becomes associated with synaptic decline and neuronal death. Because of its presence in extracellular plaques, a hypothesis was presented to describe the extracellular Aβ deposits or aggregates as the fundamental cause of neuronal death in AD.

This hypothesis states that extracellular Aβ deposits, generated by the proteolytic cleavages of APP, are the fundamental cause of AD.[34,35] As a result of its widespread acceptance and following, many publications have focused on understanding the processing pathway of APP, how it is made, enzymatic partners (β- and γ-secretase, BACE, etc.), the function and properties of its cleaved products (Aβ40, Aβ42, etc.), and how they relate to AD. Although Aβ40 and Aβ42 have been reported in plaques, the Aβ42 form is more directly toxic, has a greater propensity to aggregate, and is the most studied form of amyloid.[36] Under normal conditions, about 90% of secreted Aβ peptides are Aβ40, which is a soluble form of the peptide that only slowly converts to an insoluble β-sheet configuration and thus can be eliminated from the brain. In contrast, about 10% of secreted Aβ peptides are Aβ42, species that are highly fibrillogenic and deposited early in individuals with AD and Down's syndrome. Intracellular assembly states of Aβ are monomers, oligomers, protofibrils, and fibrils. The monomeric species are not pathological, although the nucleation-dependent fibril formation related to protein misfolding makes the Aβ toxic. The oligomeric and protofibrillar species may facilitate tau hyperphosphorylation, disruption of proteasomal and mitochondrial function, dysregulation of calcium homeostasis, synaptic failure, and cognitive dysfunction.[37,38] This hypothesis is further supported by the fact that all Down's syndrome subjects, who have the extra 21 chromosome that contains the APP gene, will have AD by the age of 40. In addition, ApoE mediates Aβ metabolism, where it can bind to Aβ to affect the deposition and clearance of Aβ. It is reported to be required for amyloid deposition in an allele-specific manner. Also, preclinical transgenic mice that express a mutant form of the human APP gene develop fibrillar amyloid plaques and AD-like pathology with spatial learning deficits. These extracellular amyloid deposits or plaques grow in size and become more toxic, eventually killing neighboring neurons and leading to AD.

The AD community's faith in this hypothesis is clearly evidenced by the funding of several highly publicized clinical trials. In many ways, the fate of AD research is contingent on the accuracy of this hypothesis and the success of the clinical trials meant to test it.

LOOKING AHEAD

Despite the number and range of attempts to understand the histopathology of AD, how the neurons die, and how to treat AD, the best current treatments can only help with the symptoms; there is no available treatment to stop or reverse the progression of AD. Advances in AD research have been challenging and without major breakthroughs in understanding its pathological basis, in spite of over 1000 clinical AD trials.[39,40] A majority of these trials have failed to deliver, for seemingly unresolved reasons.

The most recent high-profile efforts to resolve AD attempted to use an antiamyloid antibody in preclinical trials with diseased mice, with the hopes of improving memory, lucidity, and other clinical maladies in humans. Bapineuzumab, the name of this antibody, was then used in several clinical trials sponsored by several large pharmaceutical companies. The drug failed to achieve the desired end points. Even worse, in a 2012 *Forbes* article,[41] the three pharmaceutical companies announced that their Alzheimer's drug had yielded such a bad result that they were stopping all further work on it, "dashing hopes for the five million Americans suffering from AD and becoming the latest piece of evidence of the drug industry's strange gambling problem" with very high investments all into one endeavor. The article concludes that there is "no path forward" for the intravenous version of bapineuzumab in mild-to-moderate AD because of its inability to improve AD symptoms.

Perhaps the most candid statement regarding the failure of these clinical trials of the amyloid hypothesis came from Dr. Zaven Khachaturian, president of the nonprofit campaign Prevent Alzheimer's Disease 2020 and former coordinator of AD-related activities at the U.S. National Institutes of Health. In a 2014 article, he conceded,

"The amyloid hypothesis became such a strong scientific orthodoxy that it began to be accepted on the basis of faith rather than evidence … no one has stepped back to ask the fundamental question of whether our basic premise about the disease is the correct one."[42] The title of a recent PsychCentral article sums up the current state of the AD field even more succinctly: "Dead ends in Alzheimer's research expose need for new drug studies."[43] A comprehensive analysis of all AD clinical studies revealed that the failure rate of AD drug development is 99.6% for years 2002–2012, there are relatively few drugs in development, and the number of drugs has continued to decline since 2009.

So what happened? As stated eloquently by Dr. Khachaturian, the field directed all its efforts toward one approach, with abysmal results. But where are the holes in the amyloid cascade hypothesis and why did those trials fail? What can be done now to better understand the underlying mechanism of AD?

This is where my story begins. Through the circumstances of my expertise and random chance, I came into the AD field as an outsider who slowly began to unravel the amyloid cascade hypothesis. I present my findings now not to discredit the work that has brought us here today but to convince you of a fresh hypothesis of the cellular events leading to AD, and, most importantly, how to prevent it.

REFERENCES

1. Alzheimer's Foundation of America. Available at: http://www.alzfdn.org/AboutAlzheimers/lifeexpectancy.html (accessed July 30, 2014).

2. Bowen DM, Smith CB, White P, Davison AN. Neurotransmitter-related enzymes and indices of hypoxia in senile dementia and other abiotrophies. *Brain*. 1976;99(3):459–496.

3. Davies P, Maloney AJF. Selective loss of central cholinergic neurons in Alzheimer's disease. *Lancet*. 1976;2(8000):1403.

4. Perry EK, Gibson PH, Blessed G, Perry RH, Tomlinson BE. Neurotransmitter enzyme abnormalities in senile dementia. Choline acetyltransferase and glutamic acid decarboxlyase activities in necropsy brain tissue. *J Neurol Sci*. 1977;34:247–265.

5. Rylett RJ, Ball MJ, Colhuon EH. Evidence for high affinity choline transport in synaptosomes prepared from hippocampus and neocortex of patients with Alzheimer's disease. *Brain Res*. 1983;289:169–175.

6. Nilsson L, Nordberg A, Hardy JA, Wester P, Winblad B. Physostigmine restores [3H]-acetylcholine efflux from Alzheimer brain slices to normal level. *J Neural Transm*. 1986;67:275–285.

7. Whitehouse PJ, Price DL, Struble RG, Clarke A, Coyle J, Delong M. Alzheimer's disease and senile dementia: loss of neurons in basal forebrain. *Science*. 1982;215:1237–1239.

8. Bartus RT, Dean RL, Beer B, Lippa AS. The cholinergic hypothesis of geriatric memory dysfunction. *Science*. 1982;217:408–417.

9. Drachman DA, Leavitt J. Human memory and the cholinergic system. *Arch Neurol*. 1974;30:113–121.

10. Francis PT, Patmer AM, Snape M, Wilcock GK. The cholinergic hypothesis of Alzheimer's disease: a review of progress. *J Neurol Neurosurg Psychiatry*. 1999;66:137–147.

11. Moore AH, O'Banion MK. Neuroinflammation and anti-inflammatory therapy for Alzheimer's disease. *Adv Drug Deliv Rev*. 2002;54:1627–1656.

12. Krstic D, Knuesel I. Deciphering the mechanism underlying late-onset Alzheimer disease. *Nat Rev Neurol*. 2013;9:25–34.

13. Szekely CA, Zandi PP. Non-steroidal anti-inflammatory drugs and Alzheimer's disease: the epidemiological evidence. *CNS Neurol Disord Drug Targets*. 2010;9(2):132–139.

14. Mohandas E, Rajmohan V, Raghunath B. Neurobiology of Alzheimer's disease. *Indian J Psychiatry*. 2009;51(1):55–61.

15. Johnson GVW, Stoothoff WH. Tau phosphorylation in neuronal cell function and dysfunction. *J Cell Sci*. 2004;117:5721–5729.

16. Mudher M, Lovestone S. Alzheimer's disease – do tauist and baptist finally shake hands?. *Trends Neurosci*. 2002;25:22–26.

17. Iqbal K, Alonso Adel C, Chen S, et al. Tau pathology in Alzheimer disease and other tauopathies. *Biochim Biophys Acta*. 2005;1739(2–3):198–210.

18. Chun W, Johnson GV. The role of tau phosphorylation and cleavage in neuronal cell death. *Front Biosci*. 2007;12:733–756.

19. Iqbal K, Grundke-Iqbal I, Zaidi T, et al. Defective brain microtubule assembly in Alzheimer's disease. *Lancet*. 1986;2:421–426.

20. Wischik CM, Harrington CR, Storey JMD. Tau-aggregation inhibitor therapy for Alzheimer's disease. *Biochem Pharmacol*. 2014;88:529–539.

21. Giacobini E, Gold G. Alzheimer disease therapy – moving from amyloid-β to tau. *Nat Rev Neurol*. 2013;9:677–686.

22. de la Torre JC. Is Alzheimer's disease a neurodegenerative or a vascular disorder? Data, dogma, and dialectics. *Lancet Neurol*. 2004;3(5):270.

23. Lucas HR, Rifkind JM. Considering the vascular hypothesis of Alzheimer's disease: effect of copper associated amyloid on red blood cells. *Adv Exp Med Biol*. 2013;765:131–138.

24. de La Torre JC. Vascular basis of Alzheimer's pathology. *Ann N Y Acad Sci*. 2002;977:196–215.

25. www.alz.org (accessed June 30, 2014).

26. Kivipelto M, Helkala EL, Lasskso MP, et al. Midlife vascular risk factors and Alzheimer's disease in later life: longitudinal, population based study. *BMJ*. 2001;322(7300):1447–1451.

27. Hartmann T. Role of amyloid precursor protein, amyloid-beta and gamma-secretase in cholesterol maintenance. *Neurodegener Dis*. 2006;3:305–311.

28. Wellington CL. Cholesterol at the crossroads: Alzheimer's disease and lipid metabolism. *Clin Genet*. 2004;66:1–16.

29. Björkhem I, Meaney S. Brain cholesterol: long secret life behind the barrier. *Arterioscler Thromb Vasc Biol*. 2004;24:806–815.

30. Maulik M, Westaway D, Jhamandas JH, Kar S. Role of cholesterol in APP metabolism and its significance in Alzheimer's disease pathogenesis. *Mol Neurobiol*. 2013;47(1):37–63.

31. Eckert GP, Muller WE, Wood WG. Cholesterol-lowering drugs and Alzheimer's disease. *Future Lipidol.* 2007;2(4):423–432.

32. Majno G, Joris I. *Cells, Tissues, and Disease. Principle of General Pathology.* Blackwell Science; 1996.

33. Parihar MS, Brewer GJ. Amyloid-β as a modulator of synaptic plasticity. *J Alzheimers Dis.* 2010;22:741–763.

34. Hardy J, Allsop D. Amyloid deposition as the central event in the aetiology of Alzheimer's disease. *Trends Pharmacol Sci.* 1991;12(10):383–388.

35. Hardy J, Higgins GA. Alzheimer's disease: the amyloid cascade hypothesis. *Science.* 1992; 256:184–185.

36. Zhang Y, McLaughlin R, Goodyear C, LeBlanc A. Selective cytotoxicity of intracellular amyloid-b peptide 1-42 through p53 and Bax in cultured primary human neurons. *J Cell Biol.* 2002;156:519–529.

37. Mattson MP. Secreted forms of β-amyloid precursor protein modulate dendritic outgrowth and calcium responses to glutamate in cultured embryonic hippocampal neurons. *J Neurobiol.* 2004;25:439–450.

38. Pakaski M, Kalman J. Interactions between the amyloid and cholinergic mechanisms in Alzheimer's disease. *Neurochem Int.* 2008;53:103–111.

39. www.clinicaltrials.gov (accessed July 30, 2014).

40. http://en.wikipedia.org/wiki/Alzheimer's_disease (accessed August 18, 2014).

41. http://www.forbes.com/sites/matthewherper/2012/08/08/how-a-failed-alzheimers-drug-illustrates-the-drug-industrys-gambling-problem/ (accessed July 31, 2014).

42. Spinney L. The forgetting gene. *Nature.* 2014;510:26–28.

43. http://psychcentral.com/news/2014/07/07/dead-ends-in-alzheimers-research-exposes-need-for-new-drug-studies/72182.html (accessed August 3, 2014).

CHAPTER 2

Seeds of a New Perspective

… on the verge of lysing …

In order to fully understand how I arrived at this novel hypothesis, it is important to start from the beginning. In the late 1990s, I was hired by a pharmaceutical company to set up a histopathology laboratory that expanded to the Target Validation Team to support the early drug discovery programs. Simply stated, histopathology is the science of using the microscope to analyze stained tissues on slides. These techniques are typically used by pathologists in the clinical setting to view tissues under the microscope to make a diagnosis based on comparison to similar normal tissues. I use this type of work to decide whether to advance or stop drug discovery projects based on the following examples: the validation of preclinical models of human disease, the distribution on the program's target in normal and diseased tissues, and the efficacy of a drug candidate. Technically, these generally colorless tissues have to be prepared to be cut onto small microscopic slides and then stained so the scientist can visualize the tissue parts for analysis. Although the steps to prepare tissues for the microscope seem very procedural and cookbook-like, slight errors can lead to inaccurate results.

It was 1998 when I first met two of my colleagues about a project that needed my specialized histopathology expertise. I was introduced to a new research initiative to support the histopathological aspects of their Alzheimer's disease program. Having no background in AD, I immediately immersed myself in the AD literature for the upcoming weeks to familiarize myself with the state of the field and to better understand how I could best support the project.

This is when I first learned of the widely embraced amyloid cascade hypothesis of Alzheimer's disease, along with the details of amyloid, Aβ, and the especially deadly Aβ42.[1,2] The literature also identified several types of amyloid plaques, the most common being diffuse and dense-core, each based on its staining appearances. It is worth noting the established relationship between the two: the diffuse plaques become

Bursting Neurons and Fading Memories. http://dx.doi.org/10.1016/B978-0-12-801979-5.00002-3

dense-core over time as the amyloid becomes more fibrillary, eventually leading to the demise of neighboring neurons. I had not seen an AD plaque under the microscope before and depended on published papers and images for my knowledge. Many scientific papers acknowledged the Aβ42 form of amyloid as toxic and having an integral role in the onset of neuronal death leading to AD. Therefore, my starting point to help support this project was to learn how to detect Aβ42 in AD human autopsied brain tissues.

HISTOLOGY

Before these typically colorless tissues can be stained to make the amyloid plaques apparent under the microscope, the fixed tissues are first processed through routine histological methods, placed into the paraffin-wax blocks, cut into 5 μm thick slices of tissue by a microtome, and then placed onto microscopic slides for subsequent staining. Basic staining of tissues is generally like dying clothes a particular color. The commonly used stain is called hematoxylin and eosin, or simply H&E, which stains all cell nuclei purple independent of cell type and stains other structures pink such as most proteins in the cellular cytoplasm, extracellular fibers, and red blood cells (Figure 2.1).

The presence of plaques is not easily observed in Figure 2.1. To see specific parts of cells you need a method to specifically label those parts, which you cannot do with a general H&E stain. Therefore, I could not use this routine staining method to easily identify plaques. A fairly new

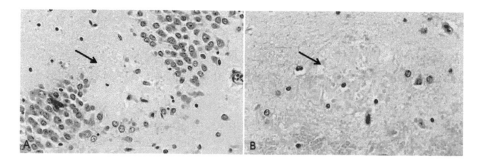

Fig. 2.1. Representative sections of the AD hippocampus (A) and AD entorhinal cortex (B) routinely stained for H&E (pink cells are red blood cells; all cell nuclei are stained purple). The presence of a plaque (arrow) in Panel A is noticeable by the loss of neuronal nuclei in the compacted layers of neurons in the hippocampus. The plaque observed in the AD entorhinal cortex is difficult to observe in Panel B (arrow) without the use of special stains (e.g., silver stain) or the technology of immunohistochemistry to detect amyloid or other plaque material.

labeling method that emerged in the 1980s and is known as immuno-histochemistry (IHC) uses antibodies to identify specific cells, cell parts, and other tissue structures that cannot be distinguished with the traditional H&E stain. These antibodies bind to their specific targets, or antigens, on the tissues just as the antibodies in your body bind to specific targets or antigens to eradicate them from your body (antigen from the word *anti*body *gen*erator). This IHC method allows one to observe specific targets on the slides with these antibodies and their specific detection system to color (e.g., brown, red, black) the presence of that antibody, which is a major advantage over the traditional staining methods that stain everything indiscriminately.

I began to explore how to perform IHC on AD tissues to observe the amyloid plaques for the first time. In order to perform the IHC necessary to observe plaques in AD tissues, I obtained a specific antibody to the Aβ42 to enable me to specifically identify these plaques.

Once I obtained the Aβ42 antibody, I performed IHC on the microscopic slides of AD brain tissues, a routine procedure for me at the time. I anxiously placed the slides one-by-one on the stage of the microscope for analysis. Just as expected, as I looked around on the slides, I was able to detect many of these reported Aβ42-positive amyloid plaques (brown-stained), both diffuse and dense-core (Figure 2.2). As a technical note, the brown staining, or labeling, is produced by the IHC detection system that indirectly, but specifically, colors the antibody to reveal the location of its target on the tissue. After adding several reagents on top of the tissue, if the antibody does not bind to its target, then the detection system will simply wash off the slide, leaving the tissue colorless. Therefore, the IHC labeling is dependent on the ability of the antibody to bind to its target.

I hope you can see the obvious benefit of this IHC method to observe plaques over the H&E staining method shown in Figure 2.1, wherein only the trained eye can see subtle differences, especially in Panel B, to discriminate the presence of plaques in the tissue. Anyone can easily identify plaques using the IHC method.

Just looking at this image of so many plaques was startling: how could someone properly function or think with these plaques in their brain? I could readily imagine neurons struggling to communicate and suffocating. I was eager to learn how my staining technique would support the

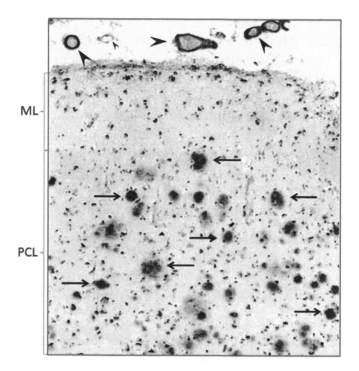

ML

PCL

Fig. 2.2. The presence of brown, Aβ42 immunolabeling in amyloid plaques (arrows) and in vascular smooth muscle cells (large arrowheads) in the AD entorhinal cortex using immunohistochemistry. Note the lack of amyloid plaques in the molecular layer (ML) of the cerebral cortex and the lack of prominent Aβ42 immunolabeling in a nearby vein (small arrowhead). ML, molecular layer; PCL, pyramidal cell layer. Further described in Biotechnic & Histochemistry. *2001;38:120–134.*

AD project, hopefully in learning how to inhibit the formation of these plaques before they kill neurons.

READY TO START

Having identified the amyloid plaques, I told my colleagues that I was now prepared to start work on the project, and we began discussing the details. These two scientists were researching a neuronal membrane receptor called the alpha 7 acetylcholine nicotinic (α7) receptor and its relation to Aβ42 in AD. The α7 receptor modulates neuronal calcium homeostasis and release of the neurotransmitter acetylcholine, two important cellular parameters involved in cognition and memory. Their idea was to block the formation of an α7 receptor:Aβ42 complex on the cell membrane, thereby reducing these amyloid plaques and possibly neuronal loss. My objective was to see if this receptor and Aβ42 were colocalized in these plaques in the AD human brain tissues. If I observed that they were colocalized, this would confirm their suspicions that blocking the receptor might stunt the development of the amyloid plaques.

I ordered an α7 receptor antibody and demonstrated its presence in the AD tissues without any issues. I was able to confirm that amyloid and this receptor were indeed colocalized in the amyloid plaques. The work was published in the year 2000 without much controversy, but what was most interesting was not the mere colocalization of this receptor and amyloid in the plaques (Figure 2.3).

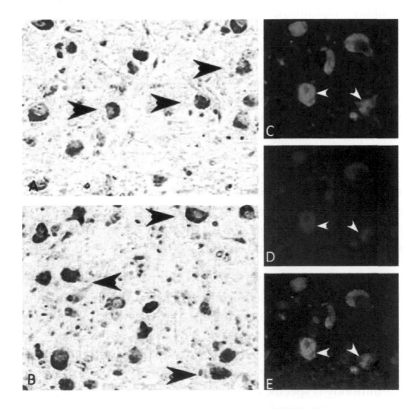

Fig. 2.3. Detection of Aβ42 (A, arrowheads) and the α7 receptor (B, arrowheads) in AD cortical neurons using IHC. (C–E) Double immunofluorescence performed on AD cortical neurons shows the presence of the α7 receptor (C, arrowheads) detected using a fluorescein isothiocyanate–conjugated secondary antibody and Aβ42 (D, arrowheads) detected using a Texas Red–conjugated secondary antibody. A triple fluorescent cube is used to detect both fluorochromes simultaneously and arrowheads indicate the colocalization or close proximity of Aβ42 and the α7 receptor, yielding a yellow color, which is the summation of the red and green fluorochrome (E). Blue 4,6-diamidino-2-phenylindole-dyed nuclei are also present. (Courtesy of J Biol Chem. 2000;275(8):5626–5632.)

I discovered something far more unexpected: besides existing in the plaques, the receptor and Aβ42 were also detected in the *neurons* (Figure 2.3)![3] Imagine my surprise to detect the supposedly toxic and hazardous Aβ42 inside the neurons as well. How were these neurons alive, and how did they come to internalize Aβ42? Also, why was the α7 receptor colocalized with the intracellular Aβ42? Shouldn't the receptor be present on the cell membrane where it normally functions?

The team continued work to show the importance of Aβ42 binding to the α7 receptor, and that this interaction could be pivotal in the pathophysiology of AD.[3] As my colleagues moved past my IHC methods, supporting these findings with nonhistological methods such as immunoprecipitation, Western blots, and biochemical methods, I began to independently investigate this surprising discovery of Aβ42 in AD neurons. It was this curiosity that ignited my passion in the AD field.

At that time, in the late 1990s, I could not find any literature to support this unique observation and I began to question if this was important to the onset of AD, a missing piece in the puzzle of neuronal death. Despite publishing this observation, I had to be absolutely sure that the IHC stains were accurate, and not just a technical artifact of the IHC method and antibody I had used to detect the Aβ42. To start my methodology check, I obtained additional AD brain sections and viewed them one-by-one under the scope. In particular, I was confirming that in the preabsorption controls (Figure 2.4), the primary antibody was specific to its antigen, in this case Aβ42.[4]

In principle, if the primary antibody is truly specific to its antigen, when you mix them together and let them incubate overnight, and then use this antibody:antigen mixture in the routine IHC methods, no free antibody should be available to bind to any of the antigen in the tissue because it is already bound to its antigen in the mixture. If no antibody binds to the tissue, then you should not produce IHC staining, thereby demonstrating that the antibody is specific to its antigen. This type of technical negative control confirms the specificity of the antibody because, conversely, if the antibody is not specific to its antigen in the mixture, then free antibody would bind to the recognizable antigens on the tissue to produce staining. Fortunately, as seen in Figure 2.4, no Aβ42 immunolabeling was detected on the tissue with this preabsorption technical control in Panel B.

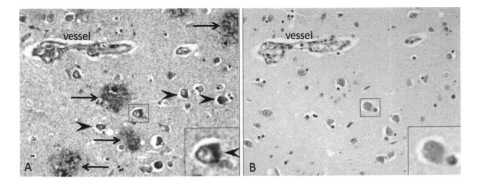

Fig. 2.4. A representative set of serial AD brain sections shows intracellular Aβ42 in neurons (A, arrowheads) and in plaques (A, arrows). The other serial section (B), as confirmed by the presence of the same vessel, shows no Aβ42 immunolabeling when Aβ42 peptide is preincubated with the Aβ42 antibody overnight prior. (Courtesy of Biotech Histochem. 2010;85(5):133–147.)

With assurance in the IHC methods, I began to carefully observe the patterns of Aβ42 in the AD tissues. I noted that the number of amyloid plaques varied across the different AD tissues of the same area in the brain, an area called the entorhinal cortex, known as one of the first areas to develop AD pathology and an area wired for short-term memory. I didn't quite know how to interpret the varying plaque numbers but was diligent in my observations all the same, confident that careful analysis of these patterns may reveal something important.

Before delving deeper into the more specific patterns that I saw, I still wanted to understand at a broader level where Aβ42 could be found in the brain. If these data suggested that the amyloid could be found within neurons, where else might it show up? My instinct was to see what I could find in non-AD brains, especially considering the detection of Aβ42 in the non-AD brain lysates using Western blotting methods.[3] To do this, I used the same IHC method to assay normal brain tissues of subjects who died from non-AD-related causes; these controls are referred to as nondemented, age-matched control tissues. I sought to compare the detection of Aβ42 in the AD brain tissues with that in the control brain tissues to further investigate the location of the labeling.

Amazingly, I did in fact observe Aβ42 in these normal neurons (Figure 2.5), but with much less intracellular Aβ42 than I detected in the neurons of the AD brains. I did not see any obvious plaques at first, but did notice an occasional plaque that seemed to be the diffuse-type,

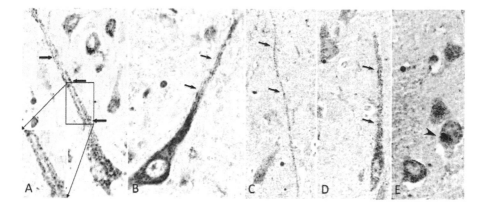

Fig. 2.5. *Representative images of Aβ42 immunolabeling in cortical sections from age-matched, nondemented controls. Note the Aβ42 distribution within the cell body and dendritic processes (arrows) and within various nearby neurons as well as the tendency for neurons to sequester their intracellular Aβ42 away from the entrance into axons and dendrites (arrowhead).* (Courtesy of *Biotech Histochem.* 2010;85(5):133–147.)

based on its "smoky" appearance. Again, the preabsorption controls indicated that the Aβ42 antibody was specific to its antigen. Recall as well that these are age-matched, nondemented control brain tissues. Despite its substantially lower intracellular density as compared with that in the AD neurons, the Aβ42 is quite apparent in the control neurons. It seemed far too coincidental for this amyloid to exist in the normal brain, let alone in the neurons, without being somehow related to the amyloid plaques associated with AD.

Now that I understood that Aβ42 was present in normal control neurons and that it was not a technical artifact, I planned to go back to the AD slides and see if I could determine the association between the Aβ42-overburdened AD neurons and Aβ42 plaques.

As presented in Figure 2.6, the amount of Aβ42 in the AD neurons varied, where some neurons had low intracellular amounts (like that observed in the age-matched control neurons) while other nearby neurons had strikingly high accumulations of intracellular Aβ42. These appeared dystrophic, or degenerative, with dense, blue-stained nuclei (Figure 2.6, arrows). In contrast, other nearby neurons with less intracellular Aβ42 had normal-appearing cell structure and nuclei (Figure 2.6).[4] Even more striking was the obvious lack of extracellular Aβ42 deposits in the forms of plaques near any of these neurons. In other words, most of the Aβ42 detected in these AD sections was present within the neurons.

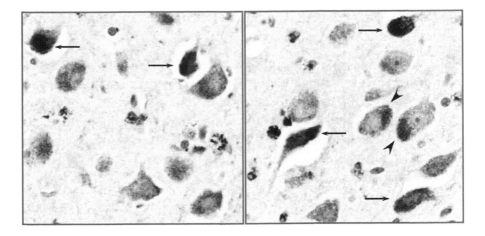

Fig. 2.6. Representative examples of Aβ42 immunolabeling in AD cortical pyramidal neurons appear to aggregate near the perikaryon in some of the neurons (arrowheads). Note also pyknotic (condensed) nuclear chromatin in nearby neurons (arrows) overburdened with intense Aβ42 immunoreactivity that is in contrast to the normally appearing nuclear chromatin with prominent nucleoli in less Aβ42-burdened neurons (bottom arrowhead). (Courtesy of *Biotech Histochem.* 2010;85(5):133–147.)

Something was not adding up about what I saw under the microscope and what I learned about AD and Aβ42. Finally, after several weeks of reviewing these and additional slides and recording countless hours under the microscope, I was forced to consider my first of many novel realizations: Aβ42 may not be toxic after all. It was clearly doing no harm in the normal brain so why should it suddenly become toxic in the AD brain, and when? What was clear was that significantly more intracellular Aβ42 was accumulating in the AD neurons. Even more apparently, the presence of Aβ42 in normal neurons in similarly aged brain tissues would suggest that the amount of intracellular Aβ42 was another indicator of the neuropathology of AD.

Perhaps there was a connection to the fact that Aβ42 was sparse in neurons in the control brains, but abundantly present in the AD neurons that appeared degenerative (dystrophic). But what kind of connection? Of course, this much intracellular amyloid could not be created from within the neuron; it would have to come from outside the neuron (perhaps entering the neuron through the α7 receptor I had investigated earlier, which might also explain its colocalization in the AD neurons). I was also seeing abundant extracellular dense-core Aβ42 plaques in the AD brain, and only rarely in the nondiseased brain. Why should that be the case? Where were the numerous dense-core Aβ42 plaques exclusive to the AD brain coming from?

An even more radical notion occurred to me: what if the Aβ42-overburdened neurons had been "consuming" extracellular Aβ42 and simply became *too* overburdened? How much amyloid could they accommodate? Could they possibly burst? It was at this point that I revisited my slides looking for evidence and after hours of analysis, I was able to locate Aβ42-overburdened neurons to support this possibility (Figure 2.7).

With some imagination, it wasn't too hard to believe that these neurons might be on the verge of lysing (the more technical term for bursting). This would explain why I could not easily find extracellular dense-core Aβ42 plaques in nondiseased brains: any extracellular Aβ42 was being ingested by the neurons, which clearly had intracellular Aβ42. But the Aβ42 wasn't toxic to them, because based on their lack of overt morphological signs of degeneration, they seemed to be clearly functioning without issue. Only when they took in *too much*, in the AD

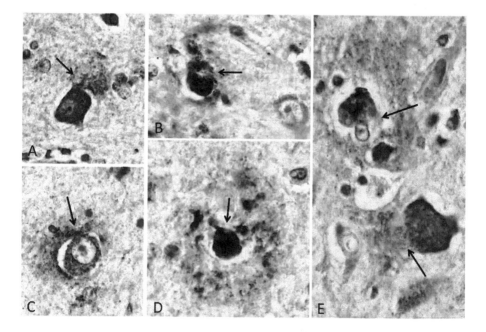

Fig. 2.7. Representative examples of Aβ42 immunolabeling in AD cortical neurons show early stages of cell lysis (arrows). (Panels C and E, courtesy of Biotech Histochem. 2010;85(5):133–147.)

brain, would they potentially lyse, leaving behind a dense-core plaque of Aβ42-insoluble residue.

Already in just several weeks and numerous slides, I had arrived at some novel provocative conclusions, while finding little morphological evidence for the amyloid hypothesis. Still, the mere presence of Aβ42 in normal neurons was not published, and there was no precedent for these findings other than my first report of the colocalization of the α7 receptor with Aβ42 in the AD neurons. I was stumped, stuck in my tracks in a field I still knew little about. I felt as though I must be missing something despite the confidence I had in my methods. I called my two AD project colleagues to my office and showed them these slides under the microscope. They seemed cautious of what I was showing them, reminding me of how unusual the implications were, and that amyloid plaques as the extracellular Aβ42 were deadly to neurons (Figure 2.3).

Despite their support of the amyloid cascade hypothesis, I still was not convinced. I decided to run a blind test. I asked an outside colleague to read these Aβ42 IHC slides, but gave no context of the slides, other than they were from AD brain tissues. For better or worse, his unbiased observations agreed with mine, which was all the reason I needed to further pursue my controversial findings.

Thus, I continued to analyze the tissues for additional morphological evidence that would further support my burgeoning hypothesis. I was able to find that additional evidence in Figure 2.8.

If you see the plaques in the images, note the presence of a nucleus in the middle of these plaques (arrows), and in particular the presence of NeuN, a neuronal-specific nuclear protein, also in the middle of a black-stained dense-core Aβ42 amyloid plaque (Figure 2.8, Panel F). For the first time, I found what appeared to be the remains of a once-intact neuron as the plaque. Another unexplained, missing piece to the amyloid puzzle, which fit surprisingly well into my solution.

In contrast to this alternative hypothesis, the amyloid hypothesis does not specify if a single plaque kills one or multiple neurons, and based on the proposed mechanism, it suggests that the neuronal nucleus of the dead neuron would be typically located on the outskirts of the extracellular Aβ42 amyloid plaque, rather than the center where I found them.

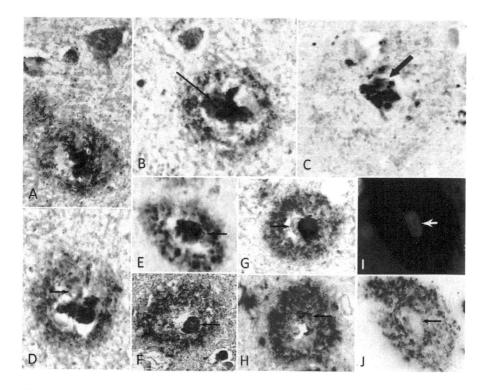

Fig. 2.8. Representative examples of Aβ42 dense-core plaques (brown A–E, J; black F; red G–H) in AD cortical tissue sections. Note the hematoxylin-stained, purple nuclei (arrows) deeply embedded in the middle of the plaques (B–H). Double IHC using a detection system to label Aβ42 black, and to label NeuN, a neuron-specific protein (arrow), red, thereby confirming the presence of a neuronal nucleus at the center of this amyloid plaque (F). Blue 4,6-diamidino-2-phenylindole-dyed nuclei (I, arrow) is observed in the center of the amyloid plaque (J). (Panels A, C–H, courtesy of *Biotech Histochem.* 2010;85(5):133–147. Panel B, courtesy of *Curr Pharm Des.* 2006;12(6):677–684. Panels I and J, courtesy of *Histopathology.* 2001;38:120–134.)

Another compelling flaw to the amyloid hypothesis was that I consistently did not observe extracellular deposits of Aβ42 contiguous with AD neurons (Figure 2.6). Most convincingly, the images in Figure 2.8 provide morphological evidence of the missing link between intracellular amyloid and amyloid plaques. Because these pivotal "action shots" were rare, they suggest that this lytic event (the "bursting") must be very fast and short-lived: it was no wonder why such a thing hadn't been reported previously.

I also took time to revisit the seemingly predictable patterns of extracellular dense-core plaques I was finding in AD tissues (recalling my observations about Figure 2.2). Before delving into what I discovered there, it is important to understand the anatomy of the entorhinal cortex. In addition to the hippocampus, an area of the brain involved in memory, the entorhinal cortex is the "ground zero" for AD and is where most of these images, including Figure 2.2, come from. The entorhinal cortex is an area responsible for short-term memory function. To best understand its anatomy as it pertains to my findings, it helps to imagine cutting a watermelon in half. Any way you cut the melon, you would see the outer crust (rind) easily distinguished from the inner fruit. Now, if you cut along the edge, you would see only the rind, no fruit.

While it is of course not so simply arranged as a fruit, the anatomy of the entorhinal cortex is well understood and easy to simplify. It is arranged in layers, from the outer (rind-like) to the inner (fruit-like) connections of the major pathways in the brain. The outer layer, also known as the molecular layer, is akin to a Brillo pad filled with cellular wiring and supporting non-neuronal cells. The "fruity" layers below house the neuronal cell bodies, which continue to increase in size as you descend through the layers of the brain. All of these neurons not only have projections that reach up to the molecular (rind) layer but also extend laterally to communicate with other nearby neurons.

Using properly oriented sections of the entorhinal cortex on the slides, I was able to draw additional conclusions relating to the behavior of plaque distributions in specific regions. First was the fact that, as can be seen in Figure 2.2, no dense-core amyloid plaques were detected in the molecular layer of the entorhinal cortex, an area that lacks pyramidal neuronal cell bodies. This suggested that the existence of neurons in a brain region determines the finding of dense-core plaques in that same region, and that the distribution of plaques among certain regions of the brain was not random, but predictable. Second, I found that the relative average size of amyloid plaques gradually increases with the size of the neighboring neuronal cell bodies as you descend down the layers of the entorhinal cortex. And third, the density of dense-core amyloid plaques increased in a specific region of the AD brain as the density of neurons decreased. This implied an inverse relationship between the number of plaques and neurons in a given region, and also supports that I rarely observed dense-core plaques adjacent to AD neurons. These data further suggested a crucial relationship between the neurons and the dense-core plaques.

The unbiased perspective provided by the IHC-stained tissues (AD and non-AD) seemed to have unlocked a wealth of new information about the histopathology of AD that incidentally provided no evidence to support the widely accepted amyloid hypothesis. I was simply overwhelmed. I had to consider one last time whether there were elements of these findings consistent with the field at the time. It seemed clear to me that I had to enter the AD field, bring these provocative findings into a public forum. I had more than just findings however. I had a seemingly coherent explanation for them, and had already independently assessed and challenged them, and found new, unexpected supporting evidence. I had a new hypothesis for the field, and I had to make it known.

REFERENCES

1. Hardy J, Allsop D. Amyloid deposition as the central event in the aetiology of Alzheimer's disease. *Trends Pharmacol Sci.* 1991;12(10):383–388.

2. Hardy J, Higgins GA. Alzheimer's disease: the amyloid cascade hypothesis. *Science.* 1992; 256:184–185.

3. Wang HY, Lee DHS, D'Andrea MR, Peterson PA, Shank R, Reitz A. β-Amyloid1-42 binds to α7 nicotinic acetylcholine receptor with high affinity: implications for Alzheimer's disease pathology. *J Biol Chem.* 2000;275(8):5626–5632.

4. D'Andrea MR, Nagele RG. Morphologically distinct types of amyloid plaques point the way to a better understanding of Alzheimer's disease pathogenesis. *Biotech Histochem.* 2010;85(2): 133–147.

CHAPTER *3*

Introducing the "Inside-Out" Hypothesis

... a logical and unifying explanation ...

There was no doubt in my mind that I was onto something important, something I could not keep to myself. I had organized these findings into a logical hypothesis, all of which I intended to present to the field.

Again, all the findings through my IHC work point to one coherent story of events of the pathology of the AD brain. These novel observations suggest that Aβ42 accumulates in neurons over time, perhaps via the α7 receptor. In the case of AD, these neurons become so engorged with Aβ42 that they die via cell lysis. This leaves the neuronal remnants in place as the Aβ42, dense-core, amyloid plaque; the "neuronal debris field" is the amyloid plaque. I have coined the term "inside-out" hypothesis to capture this suggested series of events, so named for the fact that amyloid plaques come from "inside" the neurons and burst "out" of the neurons, rather than simply existing between the neurons and eventually engulfing them. This is in stark contrast to the amyloid cascade hypothesis suggesting that neurons mysteriously die from the "outside" accumulation of extracellular amyloid. The fundamental approaches of these hypotheses are conflicting and have serious implications on the direction of AD research: the goal of my new hypothesis is to simply prevent the accumulation of Aβ42 in neurons before they die, while the goal of the current amyloid hypothesis was to simply inhibit Aβ42 from depositing outside of the neurons before they die, which failed.

I have found the simplest way to describe this new hypothesis is by way of an analogy. It is useful in this case to think of a neuron as a house. If you happen upon a row of burned out buildings, and have a primitive sense of how fires work, would you look at it and say that the residual black carbon destroyed the buildings? Probably not, because you already know that black carbon residue was the result of the burning of flammable parts in the building such as wood or paper. For decades, AD research has suggested that the "black carbon residue" (amyloid buildup) outside of the cells is the first event leading to more and

Bursting Neurons and Fading Memories. http://dx.doi.org/10.1016/B978-0-12-801979-5.00003-5

more buildup that eventually destroys the houses (the neurons). It would be as if the black carbon residue first "seeded" in the neighborhood, spread and grew by itself, and eventually consumed the houses, leaving nothing but black carbon residue in the neighborhood.

While the notion of "neuronal residue" helps to explain the faulty logic of the amyloid cascade hypothesis, the notion of a fire destroying the "neuron house" is not as accurate. For that, it is more fitting to imagine a house so completely filled with clutter that it is impossible to move around from room to room. While clutter would not lead to the loss of the house, it would make it difficult to easily maneuver. This is like a cell trying to conduct normal business where the parts need to shuffle important substances internally as well as in and out of the cell, and if it was difficult to operate, most of the normal tasks would suffer. I've seen neurons (Figure 2.5) that appeared to accumulate so much amyloid that it would be impossible to imagine the cell physically or mechanically operating normally.

Neurons dying by excessive accumulation of amyloid literally burst (lyse), leaving their contents in place. This allows unattended intracellular enzymes (the fire) to consume without control, thereby perishing lytic-sensitive cell parts (flammable items), while leaving lytic-resistant (fire-resistant parts such as metal) cell parts in place, signaling local inflammation (response of the fire engines).

The everyday person would never likely believe that the charred remnants of a home were the catalyst for the fire, and yet this is what the mainstream amyloid cascade hypothesis presumes: the amyloid started the fire between the houses that became increasingly toxic and killed neighboring neurons. Nothing can be further from what my evidence has shown.

PRESENTATION TIME

I quickly assembled all of the data into a manuscript and while my first AD paper was in review, I also submitted an abstract to the Annual International Neuroscience meeting in Miami, FL, in 1999. I should note that abstracts (summary of your work) are not peer-reviewed-like manuscripts- but are reviewed for integrity and quality, as well as for placement into the proper scientific session. Fortunately, the abstract was accepted as an oral presentation.[1] About a month before the meeting, I learned the time of my presentation and it was the first one just after lunch at 1:00 p.m. I assembled the slides for the slide projector, and then rehearsed for my first oral presentation in the AD field.

Although presentations of research findings are routine for most, this presentation was anything but usual for me, in spite of giving numerous presentations previously. I was a relative newcomer to the AD field, attempting to sway prevailing scientific consensus toward new findings in a sound and impactful manner. The day of the presentation arrived; lunch had finished and I awaited my introduction by the stage. The chairman of that session opened the meeting, and began to introduce the talk even noting his own interest in the topic and its potential implications, and then he called my name. And with that, my first step into the public domain of Alzheimer's disease research was taken.

I began by highlighting my background in immunohistochemistry, how my journey with AD research began tangentially through the support of an AD program. And as I learned more and more about the current understanding of AD, I had more and more questions about the currently accepted and highly published amyloid cascade model based on my findings, and then I asked:

> If the amyloid plaques are believed to form from extracellular deposits of Aβ42 that aggregate and grow that are toxic to neurons, then why isn't there one huge plaque in the AD brain? Why or how do they stop growing? And, why are those plaques of particular sizes and shapes, and located in particular areas of the brain? And why is Aβ42 detected in AD neurons, and more importantly, if Aβ42 is toxic, why do I detect it in normal neurons?

Then I said, "I'm about to tell you why."

And then I realized that I had everyone's attention and then felt a surge of excitement. I essentially reported the litany of compelling experimental data on why it makes more sense to believe that amyloid plaques originate from dying, amyloid-overburdened neurons, than it was to believe that amyloid aggregates and grows between neurons and, for some undefined reason, kills neighboring neurons.[2] I continued to show slides of intracellular amyloid, specifically Aβ42, in normal, nondemented, age-matched control cortical neurons and when you compare with those neurons in the AD brains, the Aβ42 in those neurons accumulates to a point that the neuron dies, leaving the dead neuron as the amyloid plaque. I also said that I had not observed any evidence to support the amyloid deposition hypothesis. In sum, I presented a logical and unifying explanation of neuronal death and plaque formation in Alzheimer's disease. I closed the talk to a nice round of applause and readied myself for questions.

I fielded a few thoughtful inquiries; most were technical questions about methods, antibodies, and whether I had tried other methods to validate those immunohistochemically based findings. However, the inevitable and frankly, exciting, moment came when I received the first criticism. A scientist came forward with a comment, stating that all of my data were suspicious and artifactual based on the cross-reactivity of cellular lipofuscin with the antibodies. Admittedly, at the time I didn't know that much about lipofuscin and my reply was a bit clumsy. I said that I appreciated his comment and said, "From what I know about lipofuscin, it is a pigment that is typically located in the neuronal perikaryon (cell body), which is around the nuclear area and I had additionally showed intracellular Aβ42 in the dendritic neuronal processes as well where lipofuscin is not typically present." Nonetheless, I was not convincing and felt a bit uncomfortable that I was not more forthright.

After another round of applause, I walked off the stage, but instead of returning to my seat, I continued to walk out the door to breathe. Just as I left the auditorium, several audience members approached me expressing their fascination. One person actually noted that he also has seen intracellular Aβ42 and his paper was also in review, but, surprisingly, he did not link the intracellular Aβ42 with plaque formation as I clearly presented. After the conference ended, I gave an interview that was published in *Science News* and summarized the presentation for the AD community at large[3]:

Michael R. D'Andrea offered the controversial idea that plaques observed in patients' brains arise from clumps of beta-amyloid in living cells. He says that the alpha 7 receptor may draw beta-amyloid into a cell until the cell dies and disintegrates, leaving an extracellular plaque. The distribution, density, and shape of plaques support this idea, he argues.

REFERENCES

1. D'Andrea MR, Nagele RG, Wang H-Y, Peterson PA, Lee DHS. Origin of an amyloid plaque in AD. Society of Neuroscience 29th Annual Meeting; 1999; Miami, FL.

2. D'Andrea MR, Nagele RG, Wang H-Y, Peterson PA, Lee DHS. Evidence that neurons accumulating amyloid can undergo lysis to form amyloid plaques in Alzheimer's disease. *Histopathology*. 2001;38(2):120–134.

3. New insight into Alzheimer's disease. *Science News*. 1999:156:319.

Addressing Technical Concerns

… is like looking at the accumulative history of the person …

An important aspect of these discoveries regards the methods and techniques utilized in the process. I dedicated an entire chapter to the technical aspects of the data because the methods are the foundation of this new hypothesis. In fact, it is the reproducible methods that are the basis of most scientific discoveries. It is the methods that generate data leading to interpretation and discovery. I remember going back to the lab after that initial presentation of my work at the International Meeting in Miami, FL (1999),[1] being puzzled by a comment that doubted the detection of Aβ42 in the neurons, implying that it has never been reported and that the intracellular labeling is an artifact due to lipofuscin background labeling, meaning that the Aβ42 antibody cross-reacted or bound to lipofuscin in the neurons giving the false-positive result of the presence of Aβ42 in neurons.

LIPOFUSCIN

Lipofuscin, often used as a histological index of aging, originates from lysosomes within the neuron and is a special type of indigestible material that gradually accumulates in long-lived cells that are terminally differentiated (do not divide) like neurons and cardiac muscle fiber cells. In unstained tissues, this pigment will appear yellow to light brown in neurons. A good analogy would be car tires in a landfill that remain for years. Unlike many other cell types of the organs in the body that continually divide, in the brain, neurons rarely multiply even though they constantly rewire their processes. Therefore, analyzing brain tissues under the microscope is like looking at the accumulative history of the person. Increases in lipofuscin above normal levels in neurons are associated with neurodegenerative diseases including AD. However, published research claimed that these increases are a natural consequence of advanced age and not related to AD. In fact, lipofuscin had been reported to produce artifacts in IHC labeling (producing false-positive staining as noted by that audience member), so I needed to develop some expertise

Bursting Neurons and Fading Memories. http://dx.doi.org/10.1016/B978-0-12-801979-5.00004-7

in this area of IHC to further support the claim that Aβ42 is present in neurons (normal and AD), in spite of my compelling technical IHC controls (Figure 2.4).

I reviewed the literature to learn of a histochemical "special" staining method back in the 1950s to stain lipofuscin red in tissue sections of brain tissue. However, the dilemma wasn't just to detect lipofuscin, but to stain the tissues to detect the lipofuscin and Aβ42 simultaneously. To do this, I needed to design a novel double staining method to stain lipofuscin red using this special staining method and to simultaneously use the IHC method to detect the Aβ42 in another color (e.g., brown or purple), so I could easily distinguish the two proteins of interest in brain tissues. If the results demonstrated that all of the Aβ42 immunolabeling was detected with lipofuscin, then there would be no way to argue against the claim that the detection of intracellular Aβ42 was an artifact.

Designing novel histological staining methods was my bailiwick, as I could design just about any type of single or double IHC staining method with another type of non-IHC staining method (e.g., van Gieson, in situ hybridization, TUNEL) to fit the needs of the study as I had previously performed for many drug discovery programs that resulted in technical publications.[2-4] This design would be to combine the lipofuscin stain with an IHC to produce two colors. First, I would prepare serial sections of age-matched, nondemented control and AD brain sectioned tissues, and then perform the following with those slides:

Slide 1: The stain to identify this lipofuscin in the neurons (to ensure the method worked)
Slide 2: IHC to detect Aβ42 (to ensure the method worked)
Slide 3: The lipofuscin stain method, and then the IHC method with Aβ42
Slide 4: The lipofuscin stain method, and then the IHC method with a negative control antibody
Slide 5: IHC method with Aβ42, and then the lipofuscin stain method
Slide 6: IHC method with a negative control antibody, and then the lipofuscin stain method

The purpose of the negative control antibody is to insure that any resulting IHC labeling was not attributed to the detection system that colors the presence of the Aβ42 antibody on the tissue. The design was

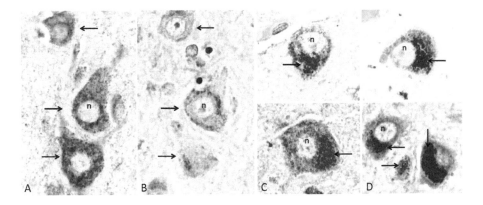

Fig. 4.1. (A and B) A representative serial set of AD cortical tissues was processed for IHC to label Aβ42 brown (A) and for HC to stain lipofuscin red (B). Three arrows indicate identical pyramidal neurons in both sections and show the presence of brown-labeled Aβ42 (A) and red-stained lipofuscin (B) in these neurons. (C and D) A novel double IHC:HC staining method was performed to immunolabel Aβ42 brown (C) or blue (D) and to simultaneously stain lipofuscin red (arrows) in age-matched control (C and D, top panels) and AD (C and D, bottom panels) cortical brain tissues. Prominent brown (C) or blue (D) Aβ42 immunlabeling is observed in the control (top panels) and in AD (bottom panels) neuronal perikaryon beyond the areas of the sequestered intracellular red-stained lipofuscin. Because there was no hematoxylin, the nuclei (n) are indicated. (Courtesy of Neurosci Lett. 2002;323:45–49.)

simple and the condition on slide #3 worked the best; in fact, the results were quite stunning. The data removed any doubt cast at the international meeting; the data were real and without lipofuscin artifact, which I published soon thereafter.[5]

Although some of the Aβ42 material seemed to colocalize with lipofuscin in neurons of the AD brain tissues, the bulk of the Aβ42 labeling was present in areas of the cells not occupied by the lipofuscin pigments in "normal, age-matched control" and AD neurons (Figure 4.1).[5]

What was most surprising to me was that lipofuscin was also detected in amyloid plaques (Figure 4.2).

Unstained lipofuscin appears as a yellow to light-brown pigment, and is clearly visible in Panels A and B (arrows) in very close association with a cell nucleus embedded in the core of the red-stained, Aβ42 plaque. Using the novel double stain to stain lipofuscin red, and Aβ42 brown (Panel C) or blue (Panel D), again shows the presence of lipofuscin embedded in the center of these plaques.

These data provoke the following question: if the amyloid plaques form from the "extracellular deposition of Aβ42," then how did the

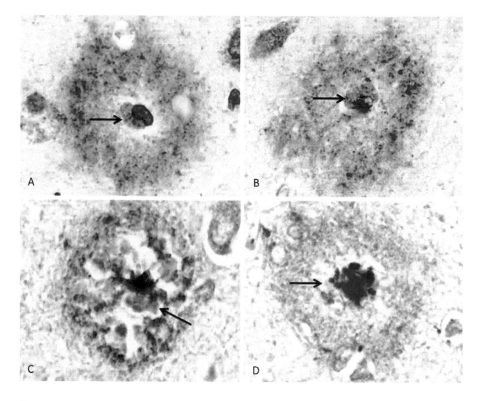

Fig. 4.2. Representative images from the AD entorhinal cortex show lipofuscin embedded deep within the middle of these amyloid plaques through various staining methods. (A and B) The presence of yellow-pigmented lipofuscin (arrows) in association with a purple, hematoxylin-stained nucleus is deeply embedded within the middle of the red-stained Aβ42 plaque. (C and D) Representative images of a double IHC-lipofuscin staining method to show red-stained lipofuscin (arrows) deep within the middle of the brown (C) or blue (D) Aβ42-positive plaque. (Panels A and B, courtesy of *Biotech Histochem.* 2010;85(2):133–147; Panels C and D, courtesy of *Neurosci Lett.* 2002;323:45–49.)

neuronal-derived lipofuscin get in there? The amyloid cascade hypothesis would predict that lipofuscin became engulfed by the enlarging (seeding) toxic amyloid plaque as the neuron dies. If this were true, I would expect lipofuscin to be on the periphery of the amyloid plaques and not in the "dead" center of the plaques, like that of the nuclei. Since lipofuscin is indigestible, as I have noted for other proteolytic-resistant neuronal cellular elements such as Aβ42, it should indeed be present in these types of plaques as a result of cell lysis.

Although this newly combined histochemical and IHC double staining method successfully ruled out possible false-positive lipofuscin staining artifacts that could affect Aβ42 detection, it serendipitously provided more compelling evidence that the dense-core type of amyloid plaque formed from neuronal lysis.

PRIMARY ANTIBODIES

One of the comments from the meeting that drove additional experimentation was the specificity of the antibodies I used in the IHC methods to detect Aβ42. I had to critically analyze the IHC method since very often if the data are unexpected or controversial, the validity of the methods is challenged. Indeed, in my 25 years in the research field utilizing IHC, I have reviewed papers with poor controls, and with the utilization of unvalidated and nonspecific antibodies, and so I am sensitive to this claim and had to address this point as well.

I needed to present a study solely based on thorough IHC details and additional antibodies, and to further address valid concerns regarding my data, I even employed a new slide-based enzyme-linked immunosorbent assay (ELISA), which is another antibody-based method to detect the presence of a particular protein in a sample. This technical paper was especially directed to those who would prefer to question the methods than to take a reflective look at the validity of the current hypothesis.[6] I remember formulating the design and even the title by considering the wording "consistent IHC detection of intracellular Aβ42 ...." Consistent is the hallmark of any claim that stems from technical data. I began my scientific career perfecting methods and even as far back as 1980, I mastered the methods of the electron microscope (EM) where all histological methods are performed under a regular light microscope before putting the tissue (specimen) in the EM. The cutting methods involve a far greater degree of difficulty than that of the histology, and I was performing IHC methods at the EM level as well, in fact, performing double immunogold methods with 5 nm gold on one side of the 3 mm 50-mesh nickel grid (akin to a slide for a regular microscope) with mounted tissue sections, and then another IHC to detect another protein using 10 nm gold particles on the other side of the grid to detect two targets simultaneously. I was confident in my technical skills and although the IHC data were challenged, I was successful in their exhaustive validation.

HEAT IS THE TICKET

Back in the late 1990s and early 2000s, those who followed cookbook recipes to perform IHC were not typically savvy to newer methods that use heat to pretreat the tissues. This would become one of the first technical issues I would address. In order for the primary antibody to reach its antigen in the formalin-fixed tissue, several methods were developed to help the antibody penetrate through the formalin cross-linking net in the tissue. Formalin is used to preserve the tissue in the natural state when removed from the body. However, many antigens, which are the targets of the primary antibodies, are inaccessible to those antibodies because of the formalin cross-linking. The use of heat through a pressure cooker, microwave, autoclave, or water bath has been reported to reverse or break the cross-links within the formalin-fixed proteins to allow the antibody to bind to its antigen (target).[7] Other non-heat antigen recovery approaches include the use of enzymes such as pepsin, or proteinase K on tissues for the same purpose. And so, I wanted to see if pretreatment methods could account for the lack of confidence of detecting Aβ42 within neurons. For this set of experiments, I used a serial set of AD tissue sections to compare no pretreatment, enzyme pretreatment, and heat pretreatment.

As presented in Figure 4.3, although intracellular Aβ42 was detected in plaques, neurons, and vascular smooth muscle cells with all three

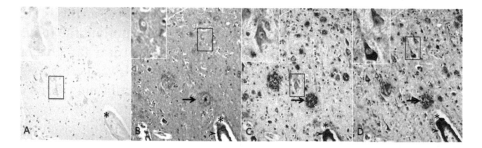

Fig. 4.3. Serial set of AD brain tissues (asterisks indicate same vessel) processed through various IHC pretreatment conditions to include heat (A, D), the enzyme pepsin (B), and no pretreatment (C). Lack of brown immunolabeling is presented in the negative control (A) to rule out potential background labeling. All three pretreatment conditions show Aβ42 immunoreactivity in the same amyloid plaque (arrows) as well as in the vascular smooth muscle cells (arrowheads). Black boxes indicate areas of magnification of the insets and show various degrees of intracellular Aβ42 immunolabeling. Further described in Neurosci Lett. *2002;333:163–166.*

conditions, the intracellular Aβ42 was most prominent using the heat pretreatment condition (Panel D). Since intracellular Aβ42 was detected using all three conditions, the pretreatment variable did not explain why other AD researchers were unable to report Aβ42 in neurons. I remained perplexed because even without a pretreatment, in spite of a cloud of light-brown background staining, I was still able to show intracellular Aβ42 in neurons and smooth muscle cells (Panel C).

ANTIBODIES

The next task was to establish the validity of the primary antibodies I used and so I obtained Aβ42 antibodies from four different vendors (Figure 4.4). The results in Figure 4.4 clearly show similar, if not identical, IHC immunolabeling patterns of Aβ42. The results of the fourth Aβ42 antibody, which is not shown, produced similar IHC labeling patterns. The conclusion was that the detection of Aβ42 in neurons was not due to issues with the Aβ42 primary antibodies since they all produced intracellular Aβ42 immunolabeling in addition to labeling the amyloid plaques.[6]

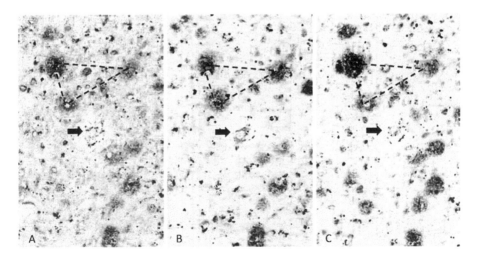

Fig. 4.4. Immunohistochemical detection of Aβ42 in serial sections of the AD entorhinal cortex (arrow indicates same vessel) using anti-Aβ42 antibodies from three unique vendors (A–C). All IHC labeling patterns appear similar as indicated by the same three amyloid plaques (arrange like a triangle). Further described in Neurosci Lett. 2002;333:163–166.

IN SITU ELISA

Lastly, I designed a novel preabsorption method as a slide-based ELISA to further demonstrate the sensitivity of those antibodies to their specific antigens.[6] The specificity of the antibodies was further verified by immobilizing 1 mM of peptides (Aβ1-42, Aβ1-43, Aβ1-40, Aβ40-1, galanin) and 0.1% proteins [ovalbumin, bovine serum albumin (BSA)] in each well of an eight-well chamber slide, rinsed, and then fixed, rinsed, and air-dried. Slides were then processed for a routine single IHC using the primary antibodies specific to Aβ40, Aβ43, and Aβ42. These data shown by the brown stain in Figure 4.5 demonstrate the antiamyloid primary antibodies to Aβ40 (slide 40), Aβ43 (slide 41), and Aβ42 (slide 42) are specific to their respective antigens. In addition, these anti-Aβ42 antibodies did not cross-react with amyloid precursor protein in Western analyses.[6]

Mind you that all of this work was an effort not only to show how to correctly and reproducibly perform IHC, for reproducibility is a hallmark of all experiments, but also to try and understand the skepticism of my data. The data were clear that the several pretreatment methods using four different primary antibodies consistently detected

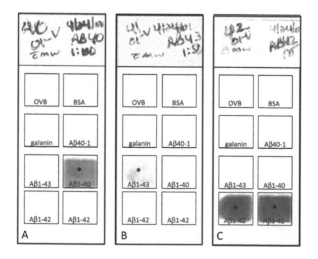

Fig. 4.5. Results of a unique immunoblotting method show that primary antibodies Aβ40 (A), Aβ43 (B), and Aβ42 (C) detect their specific peptides (1 mM) in their specific wells. Immobilized peptides are indicated by the brown reaction in their specific wells* [1 mM of Aβ1-42 (in two wells), Aβ1-43, Aβ1-40, Aβ40-1, galanin; and 0.1% ovalbumin, bovine serum albumin (BSA)]. Further described in Neurosci Lett. 2002;333:163–166.

Aβ42 in neurons. So then, why has intracellular Aβ42 not been reported? Perhaps, if you are unaware of Aβ42 in neurons and are only gauging the success of your IHC data on the detection of amyloid plaques, then how would you know you missed this very important finding?

These studies provided additional confidence in the IHC methods to consistently detect Aβ42 in neurons. The question remained: why has no one else reported similar findings?

REFERENCES

1. D'Andrea MR, Nagele RG, Wang H-Y, Peterson PA, Lee DHS. Origin of an amyloid plaque in AD. Society of Neuroscience 29th Annual Meeting; 1999; Miami, FL.

2. D'Andrea MR, Alichnavitch M, Nagele RG, Damiano BP. Applications of simultaneous PCNA- & TUNEL-labeling to evaluate testicular toxicity suggests apoptosis is more sensitive to toxicity than proliferation. *Biotech Histochem.* 2010;85:195–204.

3. D'Andrea MR, Rogahn CJ, Damiano BP, Andrade-Gordon P. A simultaneous histochemical and immunohistochemical staining protocol to evaluate 4 differently stained cell types in restenosis. *Biotech Histochem.* 1999;74(4):172–180.

4. D'Andrea MR, Foglesong PD. Simultaneous detection of protein and mRNA for DNA topoisomerase IIβ in human tumors. *Bull N J Acad Sci.* 1995;40(2):17–19.

5. D'Andrea MR, Nagele RG, Gumula NA, et al. Lipofuscin and Aβ42 exhibit distinct distribution patterns in normal and Alzheimer's disease brains. *Neurosci Lett.* 2002;323(1):45–49.

6. D'Andrea MR, Nagele RG, Wang H-Y, Lee DHS. Epitope recovery immunohistochemical methods insure intracellular localization of Aβ42 within Alzheimer's disease pyramidal neurons. *Neurosci Lett.* 2002;333(3):163–166.

7. Shi S-R, Cote RJ, Taylor CR. Antigen retrieval immunohistochemistry: past, present, and future. *J Histochem Cytochem.* 1997;45(3):327–343.

CHAPTER 5

The Good Intentions of Formic Acid

… running like mascara …

I questioned why I could not find other reports of researchers detecting intracellular Aβ42 in neurons back in the late 1990s. I continued to feel that the answer must lie in the IHC methods and the reagents. One thing for sure, I can certainly understand why scientists supported the amyloid cascade hypothesis if no one believes that Aβ42 is first in neurons. But if I could determine why scientists missed this important discovery, then others could replicate these novel findings. Perhaps its roots are embedded in the early slide staining days. Many years before IHC and after the H&E, histopathology laboratories developed staining methods, still referred to today as "special stains," to help identify tissue or cellular elements beyond the typical H&E stain that only stains nuclei purple, among a backdrop of pink without any regard to cell type (Figure 5.1).

As you can see in Figure 5.1, trying to observe amyloid plaques in an H&E slide is difficult even to the trained eye. You can see a disruption in the pattern of the row of neuronal nuclei in the AD hippocampus on Panel A, but if you look at Panel B, can you see some evidence of a plaque? The arrow shows the area of a plaque, but you can't see directly the plaque, only the impact of something that seemed to change the texture of the brain tissue. New staining methods were needed to clearly identify amyloid plaques and not simply its pathological aftermath.

These "special stains" advanced the histopathological field by analyzing more specifically stained features and cells. For example, some of these special stains could stain the muscle fiber elastin black, while other stains could stain all collagens blue. Most of these staining methods depend on the electrostatic charge and pH of the tissue or cellular elements to produce the color of choice. One such stain often used in AD research is a silver stain, and so I decided to try and stain a couple of AD tissues with this stain to understand these methods. My first impression of tissues with this stain was it needed more fine-tuning to

Bursting Neurons and Fading Memories. http://dx.doi.org/10.1016/B978-0-12-801979-5.00005-9

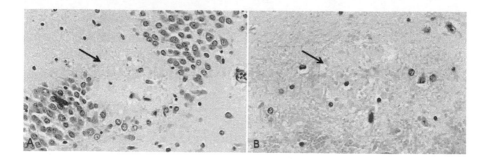

Fig. 5.1. Representative sections of the AD hippocampus (A) and AD entorhinal cortex (B) routinely stained for H&E (pink cells are red blood cells; all cell nuclei are stained purple). The presence of a plaque (arrow) in Panel A is noticeable by the loss of neuronal nuclei in the compacted layers of neurons in the hippocampus. The plaque observed in the AD entorhinal cortex is difficult to observe in Panel B (arrow) without the use of special stains (e.g., silver stain) or the technology of immunohistochemistry to detect amyloid or other plaque material. (Same figure as Figure 2.1 is reprinted in this chapter for convenience.)

visualize the tissues since they usually appeared quite messy, filled with black, metallic stained filaments, with a yellow hue all over the tissue (Figure 5.2).

If you see Figure 5.2, the plaques are easily identifiable (arrows), although trying to observe subtle intracellular details seemed more challenging to me. Therefore, I believe that since early pathologists could see the dense-core plaques, and since they appeared outside

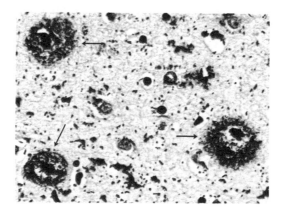

Fig. 5.2. A representative section of the AD entorhinal cortex stained using a modified Bielschowsky silver stain shows the obvious presence of plaques (arrows) among other darkly stained material that is difficult to characterize. (Courtesy of *J Biol Chem.* 2000;275(8):5626–5632.)

the cells, it's easy to understand their initial observations that these plaques must have formed outside the neurons. But then how did those early pathologists incorporate the process of neuronal death? Since there appear to be plaques of different sizes, I would think they postulated that they grow to the point where they become more and more toxic to eventually kill local neurons, which seems a reasonable conclusion given the staining technology available to pathologists of that time.

I began learning IHC in the mid-1980s when it was just being developed. I did not have experience with the special staining methods such as the silver stain because the IHC methods were much more powerful, providing unimaginable information through specific labeling characteristics using antibodies. I learned how to use IHC to identify just about anything in tissues and cells. Like many new methods, when IHC was relatively new, it was more difficult to publish manuscripts based on IHC data. Presently, IHC has a vast range of applications in research and in the clinical setting to support diagnosis and prognosis of human diseases, and publications can be found throughout all kinds of research, including AD.

However, in AD research, I learned that formic acid was a popular reagent to use in IHC methods on autopsied human brain tissues. I studied the application of formic acid and learned that it was an antigen unmasking method used to make the dense brain tissue more penetrable for the antibodies to their targets in IHC staining. This enhancement mechanism is thought to limit proteolysis and denaturation of amyloid proteins. Also, since IHCs also use a pretreatment step with enzymes or heat to overcome the formalin in the fixed tissues, again to make the tissue more penetrable to antibodies and staining reagents, perhaps formic acid could be an important variable to examine to help explain why others do not report, support, and appreciate the significance of intracellular Aβ42 in neurons.

FORMIC ACID IS THE PROBLEM

I explored the use of this formic acid on the AD tissues. Based on the literature, most methods used formic acid at room temperature. However, the concentrations and treatment times of formic acid averaged from 88% to 100% and from 5 to 60 minutes. After several test runs, 88% formic acid (pH 1.0) for 30 minutes was used in this study. After the slides were processed through the two pretreatment methods, heat and formic acid, all of the slides were identically processed using routine IHC methods to detect Aβ42.[1]

To begin, I can see why the AD research scientist routinely used formic acid to pretreat the tissues before the IHC methods to detect Aβ42 because the dense-core amyloid plaques jumped right off the slide (Figure 5.3). Many of these amyloid plaques stained so dark that you almost didn't need a microscope to observe them. They were also quite evident on the slides pretreated with heat and without formic acid, but without a doubt, if you wanted to see only amyloid plaques, the formic acid was the ticket. In terms of the quantity and morphology of the amyloid plaques, as I compared them in the serially sectioned slides, I didn't observe any numerical difference, and the morphology appeared somewhat similar, other than the intensity of the immunolabeling. The most incredible part of this

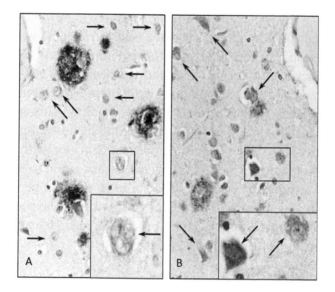

Fig. 5.3. Representative images of the AD entorhinal cortex (not serial) processed to detect Aβ42 using IHC after pretreatment with formic acid (A) or heat (B). Amyloid plaques are observed with both pretreatment methods. Importantly, intracellular Aβ42 is easily detected in the tissues pretreated with heat (B, arrows) that is clearly not observed in the tissues pretreated with formic acid (A, arrows). Further described in Neurosci Lett. 2003;342: 114–118.

experiment was about to be unfolded. Relative to the strong Aβ42 detected in the plaques using formic acid, Aβ42 was mostly absent in the AD neurons, but was easily detected in the same neurons on the adjacent serially sectioned slides using the heat IHC method. It was within this extreme divergence of detection that I believed was the potential explanation for the inability to recognize and appreciate the significance of intercellular Aβ42.[1]

How can the Aβ42 be detected in nearby neurons using one slide staining method (with heat) be completely missing using another slide staining method (with formic acid)? Could it be that the formic acid leached the amyloid out of the neurons on the slide and into the formic acid solution during the pretreatment? I continued to examine the serially stained slides to determine an explanation. I noticed that the morphology of some of the amyloid plaques on the formic acid slides had a vague teardrop shape, not quite circular, but oblong. When you looked at surrounding plaques, others had a similar shape as if they had some polarity or direction. Of course, this had to be a complete artifact as the dense-core amyloid plaques appear fairly circular in the slides pretreated with heat.

However, I wondered if this directional morphology of amyloid plaques in the formic acid-treated slides was based on how the slides were placed in the Coplin jar of formic acid. This would suggest that first, the formic acid does affect the amyloid in the tissues and second, that the amyloid in the tissues becomes somewhat mobilized based on gravity and the stained amyloid leaves the tissue in a downward direction. To test this possibility, I used a small set of sectioned slides in the Coplin jar of formic acid: the even numbered slides were placed in the Coplin jar of formic acid with the label portion of the slide up and the odd numbered slides were placed in the formic acid, upside down, with the labeled portion of the slide down.

RUNNING LIKE MASCARA

The results were simple, and stunning.[1] This teardrop-looking amyloid was an artifact based on gravity (Figure 5.4); the even numbered slide showed the amyloid moving toward the bottom of the slide, while the

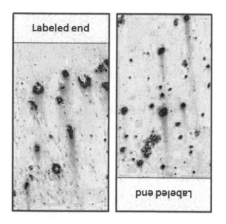

Fig. 5.4. Representative examples of the AD entorhinal cortex were simultaneously pretreated with 88% formic acid in the same Coplin jar for 30 minutes at room temperature to detect Aβ42 using IHC. The slide on the left was routinely placed in the jar with the frosted/labeling edge up; however, the slide on the right was placed in the same Coplin jar of formic acid for the same time upside down with the frosted/labeling edge down. After processing the slides for routine IHC detection of Aβ42, the results show a flow of Aβ42 downward from their representative plaques in both slides based on gravity, an artifact that was not observed in similar AD sections in the heat pretreatment condition (data not presented). Further described in Neurosci Lett. 2003;342:114–118.

next serially sectioned, odd numbered slide had the teardrop pointing in the opposite direction toward the labeled end of the slide, which was upside down in the Coplin jar. This artifact was never observed in the IHC slides I have read using the heat pretreatment.

Clearly in this simple experiment, the formic acid affects the natural state of Aβ42 in tissues. In addition to mobilizing some of the Aβ42 in the plaques, perhaps the formic acid displaced the intracellular Aβ42 out of the neurons. This possibility logically explains the lack of intracellular staining of Aβ42 in the formic acid slides producing false-negative results in the neurons. Although I could have ordered up some mass spectrometer work to verify the presence of amyloid in the formic acid solution in the Coplin jar, it was quite clear that the intracellular labeling was absent and the results offered a legitimate account why intracellular amyloid had not been detected in neurons until now.

As I prepared this publication, the initial title was "STOP using formic acid," but decided to find a title that was both compelling and informative: "The use of formic acid to embellish amyloid plaque detection in Alzheimer's disease tissues misguides key observations."[1] It was quite satisfying for me to explain how such an easy observation was missed for years. However, the use of formic acid continues to be reported in current IHC methods. The bottom line was that although the use of formic acid facilitates and embellishes the staining of plaques, it has led to conclusions based on artifact. Once again, if you evaluate IHC slides to detect Aβ42 purely based on the ability to detect amyloid plaques, then why would you question your methods suspecting the possibility of intracellular Aβ42 as a false-negative?

REFERENCE

1. D'Andrea MR, Reiser PA, Polkovitch DA, et al. The use of formic acid to embellish amyloid plaque detection in Alzheimer's disease tissues misguides key observations. *Neurosci Lett.* 2003;342(2):114–118.

CHAPTER 6

Connecting MAP-2 and Cell Lysis

… resembled Ping-Pong balls …

Another one of my continuing missions to help AD research is to assert that distinctive plaque types exist based on their origins, compositions, and morphologies. This conviction is critical in understanding and interpreting neuropathology in AD. Even after the publication of several papers, I was not able to draw enough attention to this need so I continued to provide additional validation. The data imply that some plaque types are benign or inconsequential and do not generate abnormal clinical presentations, while other plaque types evolve from neuronal death and do generate abnormal clinical behaviors. The "inside-out" hypothesis of neuronal death and dense-core plaque formation implies that clinical studies based on preclinical efficacy assessments of compounds to remove extracellular Aβ42 plaques will continue to fail because the neuron is already dead leaving no hope to reverse the damage.

The existence of more than one plaque type in the AD brain challenges the current hypothesis (amyloid cascade) that the toxic extracellular amyloid deposits aggregate to form the amyloid plaque that eventually kills nearby neurons. I challenged the supporters of the current amyloid cascade hypothesis to explain the various plaque morphologies, and why some are associated with inflammatory cells and some are not. Again, why is there not one huge plaque in the AD brain if extracellular amyloid grows? What would cause the plaque to stop growing? How does this mysterious process of deposition begin?

REVERSE LOGIC

At this point, I needed to continue to prove that the dense-core amyloid plaques originate from lysed neurons leaving proteolytically resistant cellular proteins such as Aβ42 and lipofuscin in their original place. If after the neurons dies, the cellular elements left behind are resistant to proteolysis, then the opposite must be true that proteolytically sensitive cellular proteins must be missing in the same area. I sought such a

Bursting Neurons and Fading Memories. http://dx.doi.org/10.1016/B978-0-12-801979-5.00006-0

protein and found that microtubule-associated protein 2 (MAP-2), a neuronal filament protein present in the neuronal dendrites, is sensitive to lysis (proteolysis).

I then ran serially sectioned AD and control brain sections for MAP-2 and Aβ42 IHC between serial sections to determine the relationship between Aβ42 plaques and MAP-2 labeling patterns from section to section. It is most striking when you have sections of the entorhinal cortex that clearly show the orientation of the layers from the top molecular layer down through the pyramidal cell (neurons) layers as clearly presented in Figures 2.2 and 6.1.

The results were stunning under the microscope.[1] The MAP-2 labeling was beautiful in the entorhinal cortex, as the orientation of the tissue showed parallel arrangements of MAP-2-positive dendrites, which are neuronal extensions that transmit signals back to the cell body. But what was remarkably obvious was the presence of small circular areas without MAP-2 immunolabeling that were scattered in the brain tissue resembling ping-pong balls (Figure 6.1). Also note the lack of these areas

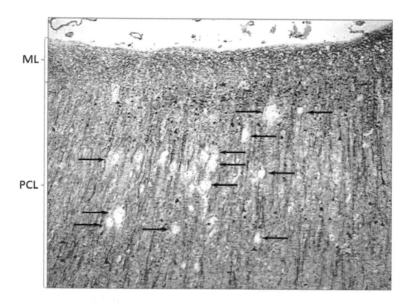

Fig. 6.1. Representative low magnification of the AD entorhinal cortex processed for IHC to detect MAP-2 that shows many areas without MAP-2 labeling (arrows) among the pyramidal cell layers that were subsequently determined to be areas of amyloid plaques. ML, molecular layer; PCL, pyramidal cell layer. Further described in Biotech Histochem. *2002;77(2):95–103.*

of MAP-2 labeling in the molecular layer, an area devoid of pyramidal neuronal cell bodies, like that previously described for the dense-core amyloid plaques.

Higher magnification of these areas clearly showed the missing circular areas of MAP-2 immunolabeling (Figure 6.2). It was very interesting to observe the abnormally intensely stained MAP-2 in nearby neurons (arrows) suggesting a cellular response to the local events in the missing area. These images appear to show the evidence of some catastrophic event, as if something exploded in these areas much like that observed in the H&E slide in Figure 2.1A. In the context of the numerous MAP-2 fibers, which by the way show an orientation to the molecular layer toward the top right, one can only imagine how many fibers of other neurons were severed by this local disturbance. Not only did the initial neuron burst, but also in its wake, collateral damage occurs, as other neurons are negatively impacted, some of which will never recover. Also keep in mind that these representative images only present pathology observed on a two-dimensional section of tissue and that the extent of the pathology reaches well beyond that plane of the tissue section.

Those areas missing the MAP-2 immunolabeling are indeed areas of dense-core amyloid plaques based on the sizes, and arrangements. The serial section of the adjacent serial section with Aβ42 labeling showed an inverse pattern that was like a negative; where the Aβ42-positive dense-core amyloid plaques were located, the MAP-2 labeling was absent (Figure 6.3). Once again, note that in the second and third serial sections from the left in Figure 6.3, the MAP-2 immunolabeling is particularly intense in those neurons just above the area of the plaque consistent with the impact of the burst neuron on adjacent neurons. I subsequently confirmed this observation through double-IHC immunolabeling in the same section.

This single result showed that the only way to explain the loss of MAP-2 immunolabeling in these plaque areas was that either the MAP-2 protein was digested as the neuron lysed or it was modified making it unrecognizable to the primary antibody. A similar consequence was observed in the H&E slide in Figure 2.1 where there was also a circular loss of staining. Regardless, MAP-2 was affected perhaps by the enzymatic activity as the neuron lysed and was unable to be detected

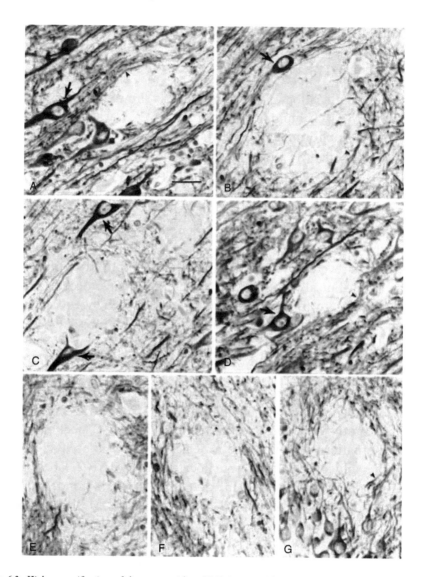

Fig. 6.2. Higher magnifications of these areas without MAP-2 immunolabeling in the Alzheimer's disease entorhinal cortex (A–D) and hippocampus (E–G) clearly show the presence of some pathological issue like that observed in the H&E stain in Figure 2.1. Note the unexplained presence of intense MAP-2 immunolabeling in neurons in the vicinity of local, MAP-2-deficient regions (arrows). (Courtesy of *Biotech Histochem.* 2002;77(2):95–103.)

in those specific areas by the MAP-2 primary antibody, which is clearly not an antibody issue as the MAP-2 immunolabeling was quite remarkable in other areas outside the plaques. As noted earlier, there have been reports of active enzymatic activity such as cathepsin D in plaques,

Fig. 6.3. Alternating serial sections of Aβ42 (A) and MAP-2 (B) immunolabeling of brain tissue from patients with Alzheimer's disease showing a spherical Aβ42-positive dense-core plaque (large arrowheads). Only one-half of the "plaque sphere" is presented in this serial set. (A) Note the presence of intracellular Aβ42-positive material in surrounding neurons (small arrowheads). (B) The missing MAP-2 immunoreactivity also occurs within a sphere that corresponds to a single dense-core amyloid plaque. (Courtesy of *Biotech Histochem.* 2002;77(2):95–103.)

specifically the dense-core types that further support the possibility that MAP-2 was digested. I further suspected that digestible parts of MAP-2 may be detectable in the CSF or blood, which could provide a biomarker of neuronal death.

DIFFUSE PLAQUES ARE BENIGN

In addition to determining that MAP-2 was missing in the dense-core plaques, I observed another surprise in the areas of the diffuse-type amyloid plaques. As their name implies, the morphology of the Aβ42 immunolabeling appeared homogeneous, as if spread out like butter on bread, without very dark or light spots of labeling, and also cloud-like. Only after comparing the serial slides for the distribution of those diffuse-type plaques on the Aβ42 slide, and then examining the same serially sectioned region on the MAP-2 slide did I realize that the normal MAP-2 immunolabeling morphology was unaffected in those areas of the diffuse plaques (Figure 6.4).[1]

Even in double-IHC slides (with Aβ42 in red, and MAP-2 in brown, see publication),[1] I could see both colors in those areas of diffuse plaques, which suggested that this type of diffuse plaque did not affect the detection or pattern of MAP-2 in the surrounding, healthy neurons.

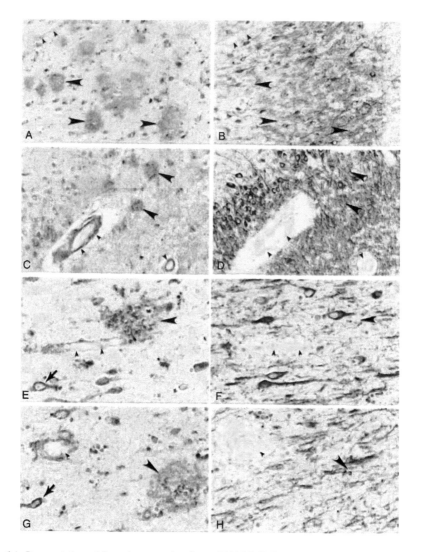

Fig. 6.4. Representative serial sets show examples of normal MAP-2 (B, D, F, H) immunolabeling (large arrowheads) in Aβ42 diffuse plaques (large arrowheads; A, C, E, G) in the entorhinal cortex (A–F) and hippocampus (G and H) of patients with Alzheimer's disease. Reference blood vessels are identified by small arrowheads in each set. Note the presence of Aβ42 in surrounding neurons (arrow) as well as in the smooth muscle cells of the blood vessels (small arrowheads). (Courtesy of *Biotech Histochem.* 2002;77(2):95–103.)

In fact, it appeared as if the presence of this "diffuse" Aβ42 was without any morphological consequence of the MAP-2 immunolabeling like a smoky cloud among rails in a banister, with the rails representing MAP-2. This is in stark contrast to the dramatic effects of the missing MAP-2 immunolabeling in the dense-core, amyloid plaques, akin to missing the rails in that banister, but this is only a two-dimensional model. Consider the negative consequences in a three-dimensional brain affecting neuronal branches from local and distant neurons, like a hole in the middle of a Brillo pad filled with electronic connections that depend on each other for communication like that presented in Figure 6.2. Furthermore, if you revisit Figure 2.1, the H&E stain in Panel A also shows the adverse effect of the bursting neuron that appears to be a hole in the row of neuronal cell nuclei in the hippocampus.

Just by performing a simple IHC to detect Aβ42 and MAP-2, I was able to clearly demonstrate the existence of at least two types of amyloid plaques: one plaque type that affected the immunolabeling of MAP-2, and the other plaque type that had no effect on the morphology of the MAP-2 immunolabeling. Actually, the loss of MAP-2 labeling may be a more effective IHC marker of neuronal death/lysis or dense-core plaques than that of the variable Aβ42 labeling.

TROUBLESOME END POINT

Much of the funding for AD research supports the amyloid hypothesis trials, which, to this day, have been unable to produce desired outcomes. If clinical studies are solely based on removing extracellular Aβ42 plaques in the AD brain, then the success of these types of clinical studies will continue to struggle no matter how early the clinical intervention. Since these diffuse plaques do not appear to be associated with cell death, these plaques are most likely unassociated with cognitive decline, and their removal should be inconsequential. Conversely, consider studies directed to remove the MAP-2-negative, dense-core amyloid plaques. Since these originate from dead neurons, why bother removing these plaques as the neuron has already died and cannot be resurrected? So, both of these approaches are fruitless, and present quite a conundrum for a field intolerant of this understanding. The removal of any types of amyloid plaques, based on the amyloid cascade hypothesis, cannot be an efficacy end point in preclinical and clinical studies and is a straightforward way to explain the failures of this approach in the clinical trials.

In spite of this reality, in order to further support or refute these possibilities, all preclinical studies need to thoroughly describe the types of plaques and intracellular Aβ42 loading and the only way to characterize them is to perform IHCs to detect MAP-2 and to detect Aβ42 without the use of formic acid. In addition, one will need to chart the behavioral changes with each plaque type change to further validate that the removal of the diffuse type of plaques will not affect behavior.

As noted earlier, it is believed that the amyloid in the diffuse plaques becomes more and more fibular, toxic, and dense over time to become the dense-core plaque, perhaps with the aid of inflammatory cells. If this were true, then one would expect inflammatory cells to be in close proximity of the diffuse plaques, but I have never seen evidence for this possibility. Therefore, I believe these are two unique plaque types with different etiologies that affect the brain differently as well.

When evaluating the data in the MAP-2 study, it was quite clear from the slides that the diffuse amyloid plaques were associated with vessels, perhaps even leaky vessels, which further suggest that the integrity of the BBB may be a contributing factor in AD pathology. This observation alone suggests that origin of these benign diffuse plaque types is from the vascular system through a dysfunctional BBB and may also explain the apparent unregulated source of the Aβ42 in neurons of the AD brain.

REFERENCE

1. D'Andrea MR, Nagele RG. MAP-2 immunolabeling can distinguish diffuse from dense-core amyloid plaques in Alzheimer's disease brains. *Biotech Histochem.* 2002;77(2):95–103.

CHAPTER 7

Classifying Plaques

... hardly recognize normal neurons ...

The data from the previous chapters have led me to dedicate an entire chapter to further explain that various plaque types exist in the AD brain because it is critical not only to the understanding of why the current amyloid hypothesis-driven clinical trials have failed but also to the overall strategy in trying to cure AD.

THREE DIMENSIONS

A challenging aspect of histopathological analyses is to account for three dimensions. For tissues with a complex organization of the cells, you should take representative slices from the tissue to ensure the presence or absence of your target before you conclude your assessment.

Let's examine another analogy, this time using a meatloaf. You cut the first slice from one end and, as expected, a slice of meatloaf with a glazing on the outer edges. As you continue to slice the meatloaf, you notice that a hard-boiled egg begins to appear with each subsequent slice, first only the white portion of the egg and then the white yolk, until the egg eventually disappears as you slice to the other end of the meatloaf (Figure 7.1).

Fig. 7.1. The meatloaf analogy of dense-core amyloid plaques in the AD brain.

Bursting Neurons and Fading Memories. http://dx.doi.org/10.1016/B978-0-12-801979-5.00007-2

To continue the comparison of slicing meatloaf to slicing AD brain tissue, in some slices you only have meatloaf without the egg, like an AD tissue without a plaque. In other portions of the meatloaf, only the white portion of the egg is visible, like the diffuse plaque, and in other portions the entire egg with the yolk is evident, like a dense-core plaque. Now to bridge the analogy to a representative set of serial slices of an AD brain, as presented in Figure 7.2, the top panels represent a single slice and are based on the immunostaining of Aβ42.[1]

The challenge when analyzing tissue sections is that you cannot faithfully produce an analysis based on a single slice; you would need to analyze serial sections to gain a more representative assessment of the tissue. In order to create a representative analysis of the true histopathology, more than one image is required, as shown in the serial strip at the bottom of Figure 7.2 that shows three panels of an identical amyloid plaque. Without taking multiple sections, the same amyloid plaque would appear as a diffuse and dense-core amyloid plaque.

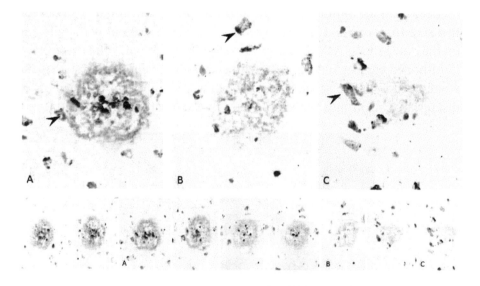

Fig. 7.2. AD tissue sections immunolabeled to detect Aβ42 illustrate the difficulty in naming amyloid plaques based solely on their amyloid morphology on a single two-dimensional histological section. Note the presence of a dense-core amyloid plaque (A), a diffuse amyloid plaque (B), and some extracellular amyloid (C). Intracellular Aβ42 is apparent in each of these sections (arrowheads). The serial section sequence shows that the appearance of a dense-core plaque in any one section depends on whether the section happens to pass through the region of the dense core. Corresponding panels (A–C) are identified in this serial sequence. (Courtesy of Biotech Histochem. 2010;85(5):133–147.)

There are several interpretations that will lead to errors: from Panel C, you may deduct that no plaques are present in the brain section; there is some extracellular Aβ42, as well as prominent intracellular labeling in nearby neurons (arrowhead). In Panel B, you would report the presence of a diffuse amyloid plaque (i.e., only seeing the white portion of the egg) as well as prominent intracellular labeling in a nearby neuron. However, in Panel A it's quite obvious to see a dense-core amyloid plaque (i.e., like seeing the white and yellow of the egg) with internal areas of fibrillar Aβ42, again among nearby neurons with intracellular Aβ42 (arrowhead). When you see the serial strip of representative sections of the same dense-core plaque, I hope you can appreciate potential for misinformation when analyzing a single section especially now understanding the shape of dense-core plaques in histological sections of tissue.

MULTIPLE COLORS

I want to use this subsection to describe IHC labeling. Most of the images are simple IHCs to show the presence of a single target (e.g., Aβ42, MAP-2, NeuN) in the tissues. However, next I want to investigate the inflammatory profile of these plaque types to further support the hypothesis that the neuronal lysis produces the dense-core plaques and activates inflammation, while the diffuse plaques do not. In order to clearly analyze plaques, I need to simultaneously detect Aβ42, microglia, and astrocytes using three unique antibodies each with exclusive and yet compatible detection color schemes. And to best explain this method, I have to take us back to the meatloaf analogy.

Consider if the meat and egg were colorless, and you are not familiar with the shape of a hard-boiled egg. Could you easily identify meat, egg white, and egg yolk? How can you distinguish the egg from the meat if both were colorless? This is the challenge of analyzing tissue sections. Stains are required to help identify structures in tissues. Additionally, research is helpful to obtain known information to understand the three-dimensional structures that are being analyzed, such as knowing that the hard-boiled egg, or in this case, the neuronal, dense-core, amyloid plaque, is spherically shaped (Figure 7.2).

If you could see the colorless slice of meatloaf under the microscope, perhaps you can see subtle differences in the morphology of the smooth egg whites from the rough morphology of the meat. If you refer back to Figure 2.1, even with the simple H&E stain, you cannot easily see amyloid plaques, but the trained eye could only notice the subtle changes in the tissue texture. IHC is a clever staining method to target specific parts of cells or proteins, and this process would make the egg show color while the meat remains colorless. To confirm the presence of meat in the section, IHC is then performed on the next "serial" slice of meatloaf with an antibody to detect meat. Using the basic IHC method, the presence of brown-stained meat in all areas of the section or slice can be observed with a hole in the middle where the colorless egg is located, which resembles the image in Figure 6.1. When you compare the two IHC-stained slides, you hypothesize that the tissue section has meat and egg whites in it, but how can you identify if egg yolk is present in the tissue? At this point, the presence of egg yolk is unbeknownst to you and you would have to take the next slice of tissue (or meatloaf) and use an antibody to detect egg yolk using the same IHC methods. But what if the slice does not contain the egg yolk and either only contains the white portion of the egg or is a slice with no egg present at all? If your slice happens to not catch the hard-boiled egg yolk in the middle, then the IHC result would be negative using an antibody to detect the egg yolk. This is the problem in trying to characterize AD amyloid plaques where three separate IHCs are performed for three different antibody targets or three serial slices of meatloaf, and deciding that only meat and egg whites are present and no egg yolk.

So there are issues in not only trying to analyze tissues in one section, which was previously described, but also how to see the meatloaf, egg yolk, and egg white at the same time, or, in this case, to see Aβ42, microglia, and astrocytes.

INFLAMMATION

Inflammation is a significant problem in the brain as the current amyloid hypothesis and many other publications suggest that extracellular amyloid is toxic and therefore elicits an inflammatory response known as gliosis. Gliosis is predominantly produced by the activity of astrocytes and microglia in the brain whose goal is to contain the injury, remove the debris, and then repair and heal the area producing a glial scar.

As noted, the next goal is to characterize amyloid plaques based on their inflammatory profile by visualizing astrocytes, microglia, and amyloid plaques simultaneously. However, there are some methodological challenges that need to be addressed. As observed in previous figures, most of the IHCs are single stains, meaning you detect a single target to produce a single color. However, in the late 1990s performing simultaneous double IHC labeling methods to see two targets simultaneously was novel but achievable as presented in Figures 2.3E and 2.8F. I could use the double IHC protocol to detect Aβ42 in red, and the astrocyte marker, glial fibrillary acidic protein (GFAP), in brown, which would allow me to see the distribution of astrocytes and Aβ42 plaques. This would enable me to analyze the astrocytes and amyloid plaques simultaneously, but I need to additionally visualize the microglia to complete the inflammatory profile.

As presented in Figure 7.3, one method to conduct this analysis would be to perform single IHC on serial sections to see amyloid, astrocytes, and microglia separately. This particular section of the AD hippocampus shows the presence of three distinct Aβ42-positive dense-core amyloid plaques (Panel A) that are associated with reactive microglia (Panel B) and activated astrocytes (Panel C). Therefore, these serial sections clearly show that these dense-core amyloid plaques are associated with inflammatory cells, but to assess hundreds of plaques, some of which would be diffuse and may not be associated with inflammatory cells, this method would require many hours to examine these individual slides under the microscope and visualize the pictures enough to make comparisons and would be prone to errors.[2] Hence, the goal was to simultaneously detect all three targets using a triple IHC with the antibody to Aβ42 to detect plaques, the antibody to GFAP to detect astrocytes, and the antibody human leukocyte antigen-D related (HLA-DR) to detect microglia cells. If I could figure out a way to perform the highly difficult assay, the stained slides could supply me a wealth of information to further characterize these plaques beyond

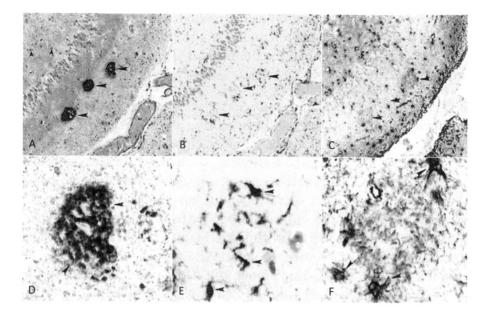

Fig. 7.3. Serial sections of the AD hippocampus tissue show the presence of brown-labeled Aβ42 (A, D), brown-labeled HLA-DR–microglia (B, E) and brown-labeled GFAP–astrocytes (C, F) and low (A–C) and high (D–F) magnifications. Arrowheads indicate the three amyloid plaques (A–C), and small arrowheads indicate the presence of Aβ42 in the surrounding pyramidal neurons (D), reactive microglia (E), and activated astrocytes (F). The serial sectioning of three separate IHCs suggests that these amyloid plaques are associated with activated astrocytes and reactive microglia (gliosis). (Courtesy of *Biotech Histochem.* 2010;85(5):133–147.)

the simple morphology of the Aβ42 immunolabeling (e.g., diffuse or dense-core). However, the challenge is finding three unique and compatible detection reagents that would not cross-react to blur the markers on the slides.

TRIPLE IHC

I envisioned a first of its kind triple IHC stain to observe Aβ42 plaques in red, among GFAP-positive activated astrocytes in black, and HLA-DR-positive reactive microglia in brown or purple so I could simultaneously characterize the association of inflammatory cells with the amyloid plaques.[2] First, I had to determine the best order of the IHC detection systems, technically to use the horse-radish peroxidase system with the black chromogen, the alkaline phosphatase system for the red chromogen, and another horseradish peroxidase system for the purple chromogen. These pilot studies were imperative to test the cross-reactivity of these particular chromogens to see if they might affect each other, especially since one of the detection chromogens was water soluble and another was water insoluble – hence, many conditions to consider that are listed. I tested these detection systems on other tissues just to ensure reagent compatibility (i.e., I wasn't going to produce mud).

1. Red → black → purple
2. Red → purple → black
3. Black → red → purple
4. Black → purple → red
5. Purple → red → black
6. Purple → black → red

Some of these combinations failed, resulting in one or two colors, but not three separate colors. The optimal combination was option #6, so I arranged the sequence order for the next set of test slides to further ensure the negative controls would be clean and to move onto AD tissues.

There were a large number of technical controls I had to assemble in order to properly assure that the detection reagents alone (i.e., the negative control: without the antibody) would not affect the other chromogens. The conditions are listed as follows:

1. Negative control (purple) → negative control (black) → negative control (red)
2. HLA-DR (purple) → GFAP (black) → Aβ42 (red)
3. HLA-DR (purple) → GFAP (black) → negative control (red)
4. HLA-DR (purple) → negative control (black) → Aβ42 (red)
5. HLA-DR (purple) → negative control (black) → negative control (red)
6. Negative control (purple) → GFAP (black) → Aβ42 (red)
7. Negative control (purple) → GFAP (black) → negative control (red)
8. Negative control (purple) → negative control (black) → Aβ42 (red)
9. Negative control (purple) → GFAP (black) → negative control (red)

I remember sitting down at the microscope to begin the analyses as I first read the three negative control conditions (#1), and it was completely clean, no labeling with colorless tissue, which is what you would expect to see in the triple negative control slide. Also, I did not see background staining on the tissues, which was fantastic. I then meticulously evaluated the single antibody slides with the two negative controls (conditions #5, #8, and #9) to ensure there was no background residue from the other chromogens, and they looked like regular single IHC slides. I only observed purple microglia in condition #5, only red Aβ42 in condition #8, and only black astrocytes in condition #9, and they were amazingly clean as well. Next, I analyzed the slides with two antibodies + 1 negative control, which looked precisely as I intended, and I jumped right to the triple IHC slide (condition #2).

Then I placed the triple IHC slide under the microscope and technically, scientifically, and emotionally, it was breathtaking (Figure 7.4).[2] I was in awe reading the slide and fortunately I did a couple of these full triples on a few other AD and control brain tissues at the same time. For the first time, I or anyone for that matter was able to see an AD brain stained in such a way to simultaneously see all three targets: I can now see the red amyloid plaques among areas of inflammation as presented by black stained, activated astrocytes and purple stained, reactive microglia.

Fig. 7.4. Triple immunohistochemistry detecting red-labeled Aβ42, purple-labeled HLA-DR-positive, reactive microglia, and black-labeled GFAP-positive activated astrocytes in a representative section of the AD brain. (Courtesy of *Biotech Histochem.* 2010;85(5):133–147.)

I recall reflecting on the individual whose life was claimed by this dreadful disease. It was apparent that this person must have suffered greatly under this terrible disease, likely unable to think clearly or perhaps hang on to memory for any nominal amount of time. Many areas of the brain were covered in plaques and many, if not most, of the plaques were shrouded with glial scars that reminded me of what a pilot might have observed days after a mission when flying over a heavily bombed area with demolished buildings amidst rubble and destroyed bridges with faint outlines of streets. In some areas of the brain, I could hardly recognize normal neurons; it was frighteningly horrible. At one time, I could feel my eyes fill up feeling so much sorrow and compassion for the donor. It was just devastating. And yet, other distant areas of the brain were totally unaffected. Looking at these sections of brain was like looking at the person's past as those neurons had been there since birth, and those dead neurons or plaques, in the context of the inflammation, were visual proof of the horrific effects of the bursting neurons leading to erased memories and function.

If you have ever used a microscope, then you know that as you increase the magnification by rotating the objective lens, and as the viewing field gets larger, it will take more and more time to scan the entire tissue section. Well, I was up to a $60\times$ dry lens objective carefully visualizing and recording each nuance, while noting the location of the purple-stained reactive microglia in relation to these red-stained, amyloid plaques while also noting the morphology of the black-stained fibers of the activated astrocytes. I could have written a book just on these slides. They were so impressive and yet so overpowering.

For those who read AD slides, I strongly urge you to triple stain your slides and then allow at least 3–4 hours to read each slide, and believe me, it will change your life as you will never read an AD slide the same. Although I cannot provide an exact measure on the amount of time spent, I intensely characterized over 100 plaques per brain section in about 50 AD brain tissues, and 7 control brain tissues (mostly hippocampus or entorhinal cortex areas of the brain). At higher magnification, many of the plaques were covered with invading black GFAP–astrocytic fibers, and three to five darkly stained purple HLA-DR–microglia cells that seems to be centrally located. The black fibers were coming in from the anchored astrocytes arranged on the periphery of the dead neuron, as if to give aid, to help fill in the empty area as a scar, but without moving. In fact, the other side of the astrocyte was acting as if nothing happened; it was quite clear this was a polar event, reacting to one side only. The control brains had little areas of inflammation, and the occasional diffuse-type amyloid plaque appeared free of reacting inflammatory cells.

MICROGLIA'S UNANTICIPATED LOCATION

The arrangements of the purple-stained, reactive microglia were surprising (Figure 7.5).[1] Most, if not all, of the literature states that Aβ42 is toxic and leads to the death of neurons. If this were the case, then why were the microglia typically found in the middle of those dense-core plaques? Why shouldn't these microglia be arranged on the periphery of the red-stained plaque like a decorative fence around the perimeter of the plaque to begin to digest, or ingest, that amyloid at first contact? I certainly would have hypothesized that based on the literature.

I can explain why the microglia are typically located in the middle and not on the periphery of the dense-core plaques. With the "inside-out" hypothesis describing neuronal death and dense-core plaque formation,

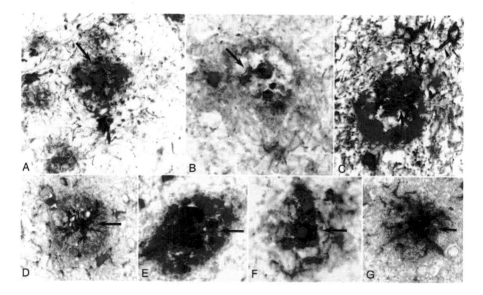

Fig. 7.5. Representative images of triple immunohistochemical labeling in AD cortical tissues showing the presence of purple-stained, HLA-DR-positive, reactive microglia (arrows) in the middle of these Aβ42-positive (red) amyloid plaques. Black-labeled (A, C) or brown-labeled (B, D, F) GFAP-activated astrocytes are observed. (Courtesy of *Biotech Histochem.* 2010;85(5):133–147.)

the neuron accumulates Aβ42 until it disturbs normal functioning leading to cell death and then lyses as its membrane collapses. So what remains in the middle of this plaque type? It is the neuronal nucleus, and its DNA, as well as any neuronal component that was resistant to the released cellular enzymes as the cell died: they remain in place at the scene of the crime.

I continued to pursue the idea that microglia are more attracted to the embedded and lysed debris of the dead nuclei than that of the released extracellular amyloid. I performed double IHCs with a newly available antibody that detects a neuronal-specific nuclear antibody protein (NeuN). The panels in Figure 7.6 show for the first time the intimate association of purple-stained, reactive microglia (arrows) with red-stained, neuronal nuclei (arrowheads), which can only occur if the nuclei are extracellular, thereby providing additional evidence about the apparent affinity of microglia to the neuronal nucleus.[1]

Microglia appear to be attracted to the core of dense-core plaques, but not to diffuse plaques, most likely by the release of DNA fragments from the nuclear remnant. I believe that the released ATP or ADP induces microglial chemotaxis via the Gi/o-coupled P2Y receptors[1,3] toward the center of these plaque types. Therefore, the role of the microglia would be to ingest such nuclear debris. On further research I learned that the microglia also have receptors such as P2X, and the scavenger receptor A (CD36) on their membrane that are activated by purines or fragmented DNA that cause the microglia to be chemoattractant, further supporting

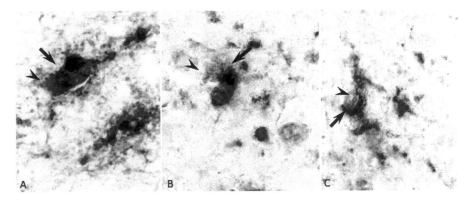

Fig. 7.6. *Representative images of the AD cortical tissues processed for a double immunohistochemical assay show reactive microglia in purple (arrows) in close association with red-labeled NeuN-positive neurons (arrowheads).* (Courtesy of *Biotech Histochem.* 2010;85(5):133–147.)

this hypothesis. Consider the ramifications of extracellular DNA. This debris could potentiate a response similar to that of an autoimmune disorder such as lupus. Since the body prioritizes the resources in an immune response, it is possible that DNA and other nuclear components that are typically located in a cell might illicit immediate concern once they are released as the cell dies.

Interestingly, I also observed red-stained, Aβ42-postive plaques with only microglia and without those black-GFAP-positive fibers, although they were not as abundant. Profiling this plaque type based on the association of inflammatory cells suggests the invading microglia are the first responders after the neuron dies, a phenomenon that can be clearly observed only in the findings of a triple IHC-labeled slide.

THE DIFFUSE AMYLOID PLAQUE

Of particular interest was analyzing the inflammatory profile of the diffuse amyloid plaques. Simply stated, they were without signs of gliosis: activated astrocytes or invading reactive microglia.[1] Of course, this was in complete contrast to the reports that all extracellular amyloid, in particular fibrillar Aβ42, is toxic.

Consider the following: since varying amounts of Aβ42 appear to be present throughout life, why would our inflammatory system be primed to attack a target that appears normal? But the elicitation of an immune response to extracellular DNA, which is always found inside the cell, follows a more logical pathway.

These data provided more compelling evidence to declare the existence of more than one plaque type in the AD brain based on the MAP-2 immunolabeling, and now on the inflammatory profile. Since these diffuse plaques did not affect local areas of MAP-2 labeling, nor do they seem to trigger inflammation, these types of plaques must be inconsequential or benign in nature.

Successful triple IHC depends on prudent selection of chromogen systems, optimization of detection sensitivity, resolution of reagent compatibility, and production of a clear signal with elimination of background to provide a rich characterization of the inflammatory profile of amyloid plaques. I encourage all AD researchers who investigate the histopathology of preclinical models of AD and those who read AD slides to assay your tissues with this triple IHC.

DIFFUSE–VASCULAR AMYLOID PLAQUES

In revisiting the meatloaf analogy, how could I rule out egg yolk from a slice that only shows egg white? In other words, how can I rule out a diffuse amyloid plaque from a dense-core inflammatory plaque? Easy, I know that gliosis encapsulates the dense-core plaque like a fibrous capsule encapsulates a tumor. If I observed a diffuse plaque with gliosis, then I could believe that if I sectioned the tissue a bit more, I will see the egg yolk or dense-core; conversely, if I observed a diffuse plaque without gliosis, I could feel a bit more confident that it is not a dense-core plaque, and therefore must be a diffuse plaque. But what is the origin of the typical, noninflammatory diffuse amyloid plaque?

As presented in Figure 7.7, the evidence is convincing that diffuse plaques originate from leaks in the vasculature. If only one section is

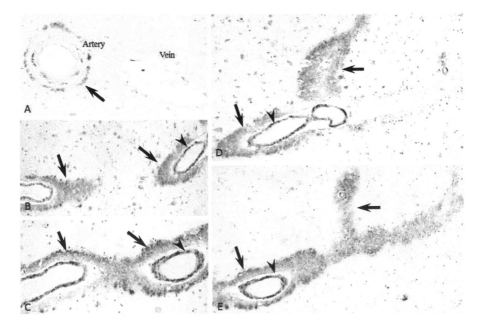

Fig. 7.7. Aβ42-positive, vascular, diffuse plaques associated with an artery (arrow) and but not with a vein (A). Vascular-associated Aβ42 resembling diffuse plaques (arrows) in serial sectioning sets (B and C, D and E) clearly show its association with the vessels (Panels B–E). Also note detection of intracellular Aβ42 in vascular smooth muscle cells (arrowheads). (Courtesy of Biotech Histochem. 2010;85(5):133–147.)

analyzed, one would infer that (1) there is extracellular Aβ42, (2) it has a homogeneous staining pattern, meaning no dense areas, and (3) it seems to be associated with a vessel. But when you perform serial sections of single IHCs to detect Aβ42, it is very clear that this type of amyloid plaque is associated with the arterial system, and not the venous system.[1] Furthermore, arrowheads show the presence of intracellular Aβ42 in smooth muscle cells, and the lack of amyloid around the vein, which does not have smooth muscle cells. These types of vascular-derived, diffuse plaques are not associated with inflammatory cells and, as previously demonstrated, do not affect MAP-2 immunolabeling either. But what is the source of this amyloid? Clearly, it is coming from the vasculature from a dysfunctional BBB.

DENSE–VASCULAR AMYLOID PLAQUES AND OTHERS

As presented in Figure 7.8, here is another set of morphologically and etiologically unique plaques. Besides the obvious presence of intracellular Aβ42 in nearby neurons (arrowheads), what is impressive is the lack of occlusion in these smaller vessels (arrows) in spite of the prominent Aβ42-immunolabeling surrounding these smaller vessels in

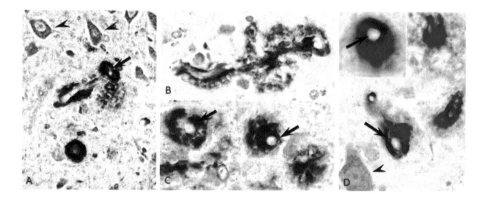

Fig. 7.8. Representative examples of intensely Aβ42-positive deposits (arrows) are associated with smaller vessels (A–C) on the AD cortical tissues. Although the amyloid is dense, the lumen of these small vessels does not appear to be occluded (arrows). Double IHC labeling confirms the presence of the red-stained Aβ42 plaques in association with the brown-labeled collagen IV-positive vessels (arrows) in the AD cortical tissue. Note the detection of intracellular Aβ42 in nearby neurons (arrowheads) independent of the detection system (brown in Panel A, or red in Panel D). (Courtesy of Biotech Histochem. 2010;85(5):133–147.)

the AD brain.[1] It's unclear how these smaller dense–vascular amyloid plaques (also can be described as amyloidosis) form in these vessels. If there are arterioles, then dying smooth muscle cells ladened with Aβ42 could play a role, but if they are capillaries, another unknown mechanism is to blame. Also other plaque types have been characterized such as those that originate from astrocytes[4] and from Purkinje cells of the cerebellum.[5]

CALL FOR A NOMENCLATURE

Precise nomenclature is paramount for any specialty, like that to address the many types of amyloid. In this example, it is important to properly identify plaques not entirely based on subjective labeling characteristics, as this can vary depending on the quality of the stain as well as the level of expertise of the investigator. To continue to name plaques solely based on their morphological appearance of the Aβ42 labeling, one would have to begin by standardizing all IHC methods, beginning with the replacement of formic acid for heat pretreatment. Maybe this is a good place to share with you other plaque names that have been reported in the literature: senile, star, monster, primitive, neuritic, classical, vascular, mature and immature, burned out, and compact and cotton wool, which, of course, are also dependent on those staining methods.

Also, there would have to be an appreciation for histological architecture; serial sections would allow one to say a plaque is diffuse if its morphology is planar or flat, not just homogeneous as this could also represent the grazing section of a dense-core plaque. Next, one would have to assess the MAP-2 and inflammatory profile of the plaques, since those from lysed neurons lack MAP-2 immunolabeling and are associated with inflammatory cells. Only when plaques are properly classified can their linkage to clinical manifestations of AD be validated.

Even the use of the word "plaque" is a concern for me. What does the word plaque convey to you? This word differs depending on the field of study. For those people in the AD field, "plaque" translates to an AD marker. But to dentists, "plaque" is the buildup of bacteria that colonize on your teeth, and for cardiovascular medical doctors "plaque" is the buildup of cells around the vascular wall, such as atherosclerosis. Based on the conventional hypothesis, buildup of amyloid nicely describes a plaque in AD. However, only the diffuse amyloid plaque fits this loose description.

How would you label the dense-core plaques based on this new hypothesis? Would you simply rename them "dead neurons"? Why even use the word plaque? "Neuronal plaque" sounds most appropriate, similar to one of the many previous names, neuritic, because as I already described, there are also glial plaques, and vascular smooth muscle plaques. Neuronal plaques contain proteolytically resistant neuronal proteins such as cathepsin D, ubiquitin, tau, lipofuscin, $A\beta42$, and DNA associated are with reactive microglia and activated astrocytes, and are missing proteolytic-sensitive neuronal proteins such as MAP-2. Diffuse plaques contain $A\beta42$. Astrocyte plaques contain the dead astrocyte and its cellular remnants such as GFAP as well as the internalized $A\beta42$, and may be associated with reactive microglia. Vascular plaques contain the dead smooth muscle cell and its cellular remnants such as smooth muscle actin and $A\beta42$ and may be associated with gliosis. And, as reported, there are Purkinje and glial plaque types in the AD cerebellum. A generic comment across all of these AD plaque types is if the cell dies, then inflammation and gliosis will be present.

Unfortunately, there are no shortcuts to effectively name a plaque and one has to characterize that plaque as you would characterize cancer through thorough IHC analyses. For example, is a particular lung biopsy positive for the epithelial marker, cytokeratin, or a melanoma marker, HMB-45? For if it is positive for the latter, then you have identified a metastatic melanoma lesion. This kind of approach must be applied when characterizing plaques, such as the use of MAP-2 as a marker of neuronal dense-core plaques. If your IHC methods are basic and you are not able to perform triple IHCs, then I would suggest you perform a simple MAP-2 IHC to categorize amyloid plaques based on their MAP-2 profile as the loss of MAP-2 immunoreactivity in the brain correlates with the presence of dense-core, inflammatory plaques.

REFERENCES

1. D'Andrea MR, Nagele RG. Morphologically distinct types of amyloid plaques point the way to a better understanding of Alzheimer's disease pathogenesis. *Biotech Histochem.* 2010; 85(2):133–147.

2. D'Andrea MR, Reiser PA, Gumula NA, Hertzog BM, Andrade-Gordon P. Application of triple-label immunohistochemistry to characterize inflammation in Alzheimer's disease brains. *Biotech Histochem.* 2001;76(2):97–106.

3. D'Andrea MR, Cole GM, Ard MD. The microglial phagocytic role with specific plaque types in the Alzheimer disease brain. *Neurobiol Aging.* 2004;25(5):675–683.

4. Nagele RG, D'Andrea MR, Lee H, Venkataraman V, Wang H-Y. Evidence for glial amyloid plaques in Alzheimer's disease. *Brain Res.* 2003;971(2):197–209.

5. Wang H-Y, D'Andrea MR, Nagele RG. Cerebellar diffuse amyloid plaques are derived from dendritic Aβ42 accumulations in Purkinje cells. *Neurobiol Aging.* 2002;23(2):213–223.

CHAPTER 8

When Is a Star Like a Plaque?

... the advent of spectroscopy and immunohistochemistry ...

A supernova is the remnant of an exploded star. This once contentious and counterintuitive fact is now a commonplace truth. How did researchers discover the correct causal relationship and not the reverse, that a supernova was formulating a star? The story had a surprising amount of parallels to my own scientific discoveries. In briefly sharing the history of stars and supernova, I hope to better elucidate my own story of neurons and plaques.

SIMILAR INITIAL HYPOTHESES

Coincidentally, the initial hypothesis about the "extrastellar" gases of supernovae was identical to the current hypothesis about the extracellular amyloid of diffuse type plaques: both eventually coalesce to form the star and the dense-core amyloid plaque, respectively.

Historically, the origin of the supernova was considered obscure despite records documenting their existence over 1000 years ago. Observers slowly came to recognize a class of stars that undergo long-term periodic fluctuations in luminosity that begin bright and then fade over months. One of the earliest purported explanations came from Tycho Brahe, in 1657, who believed that the cosmic vapors of the Milky Way coalesced to become a luminous new star, which he published in the book called *De Stella Nova* (*On a New Star*), thus coining the term "nova."[1] Another subsequent hypothesis suggested that a nova was the result of an intrinsically bright star moving toward, then away from, the Earth. Yet another proposed that the brightening of a star was due to magma seeping out through cracks in its surface.[2] In 1713, Isaac Newton proposed that novae are burnt-out stars that brightened when impacted by comets. Many subsequent theories suggested collisions with meteor streams, comets, asteroids, planets, and other stars.[2,3]

Bursting Neurons and Fading Memories. http://dx.doi.org/10.1016/B978-0-12-801979-5.00008-4

NEW TECHNOLOGY

None of these theories were readily testable within the limitations of the telescope. This all changed with the application of a new method to characterize the elemental composition of light. This new method, known as spectroscopy, finally made it possible to understand the origin of novae. It was thus the advent of spectroscopy in the second half of the nineteenth century when new theories were developed.

In 1866, English astronomer William Huggins was the first to put this new technology to use. Huggins made the first spectroscopic observations of a nova, the recurrent nova T Coronae Borealis, discovering not the broad range of lights that he expected but only bright lines of hydrogen. Huggins and his colleague William Miller were forced to conclude, for the first time, that the existence of these bright hydrogen lines was the result of a cataclysmic stellar explosion, which drew interest from other astronomers.[4,5]

REFLECTIONS OF A DISCOVERY

Much like I am trying to do here and now, William Huggins wrote a retrospective essay on his groundbreaking findings, titled *The New Astronomy*, in 1897, long after his initial discovery. In it he wrote: "The riddle of the nebulae was solved. The answer had come to us in the light itself ... I was fortunate in the early autumn of ... 1864, to begin some observations in a region hitherto unexplored." He continued, "The reader may now be able to picture to himself to some extent the feeling of excited suspense, mingled with a degree of awe, with which, after a few moments of hesitation, I put my eye to the spectroscope. Was I not about to look into a secret place of creation?"[4,6]

Huggins further described his discovery: "I looked into the spectroscope. No spectrum such as I expected! A single bright line only! At first I suspected some displacement of the prism, and that I was looking at a reflection of the illuminated slit from one of its faces. This thought was scarcely more than momentary; then the true interpretation flashed upon me. The light of the nebula was monochromatic ... and the riddle of the nebulae was solved."[6]

Huggins was bewildered by the spectrum's monochromatic character. Closer examination revealed not one, but three widely separated lines, each with its own color. The spectra of the other planetary bodies he examined shared this remarkable appearance. Furthermore, William H. Pickering later found evidence in the spectra of the bright nova GK Persei (1901) that collision theories were not valid physical explanations of novae.

REINFORCING EVIDENCE

In the aftermath of the Second World War, renowned English astronomer Fred Hoyle (who later coined "the big bang") worked on the problem of how the various observed elements in the universe were produced. In 1946, he was the first to propose that a massive star could generate the necessary thermonuclear reactions to create heavy metals, and that those reactions were responsible for the removal of energy necessary for a gravitational collapse to occur. The collapsing star would become rotationally unstable, producing an explosive expulsion of elements that were distributed into interstellar space.[7] This process of star death is of course accepted as fact today. The concept that rapid nuclear fusion was the source of energy for a supernova explosion was further developed by Hoyle and William Fowler during the 1960s.[8]

Supernovae mark the violent end to the evolution of a massive star during which the star blows out as much as 10 solar masses of material at very high speeds.[2] In the words of the legendary Carl Sagan, "Atoms synthesized in the interiors of stars are commonly returned to the interstellar gas. Red giants find their outer atmospheres blowing away into space; planetary nebulae are the final stages of Sun-like stars blowing their tops."[9] In other words, supernovae violently eject much of their stellar mass into space. The atoms in the interstellar gas left behind are, naturally, those most readily made in the thermonuclear reactions in stellar interiors. Some of the more rare elements are generated in the supernova explosion itself. We have relatively abundant gold and uranium on Earth only because many supernova explosions had occurred just before the solar system formed.[9]

To summarize, these observations through spectroscopy led to the discovery that some nebulae are formed as the result of supernova explosions, the death throes of massive, short-lived stars. The materials thrown off from the supernova explosion are ionized by the energy. One of the best examples of this is the Crab Nebula, in Taurus, that was once a giant star, about 10 times the mass of our Sun, until it exploded in a supernova leaving the collapsed star, or pulsar, in its center.

STAR NOMENCLATURE

One final note about the history of supernova bears explaining. The discovery of a new star in the Andromeda nebula in 1885 first necessitated the distinction between novae and supernovae. Walter Baade and Fritz Zwicky classified "supernova" many years later, in 1934, as explosions on the scale of 10,000 times larger than that of novae.[2,10] Baade and Zwicky performed early work on this new category of nova at Mount Wilson Observatory during the 1930s.[10] They identified S Andromedae, which they considered a typical supernova, as an explosive event that released radiation approximately equal to the Sun's total energy output for 107 years. They decided to call this new class of cataclysmic variables super-novae, and postulated that the energy was generated by the gravitational collapse of ordinary stars into neutron stars.[11] The name super-novae was first used in a 1931 lecture at Caltech by Zwicky, and then used publicly in 1933 at a meeting of the American Physical Society.[12]

SCIENTIFIC PARALLELS

I hope by now you are seeing some of the uncanny parallels that the scientific history of supernovae shares with the origin of amyloid plaques. Perhaps the most striking similarity is in the physical nature of the phenomenon itself. The image of the supernova remnant Cassiopeia A almost mirrors the image of neuronal remnant dense-core amyloid plaque (Figure 8.1, and cover).

Obviously the magnifications are opposite but keep in mind that they are "static" images from which one draws interpretation. In reference to "static images," William Hershel was the first person to understand that when looking through a telescope, "we cannot look up out into space without seeing back in time." I would expect any histopathologist to appreciate the same: when looking down through the microscope to see brain tissue, you are also looking at a snapshot representing the history of that person. However, it is in comparison to brains of other ages and

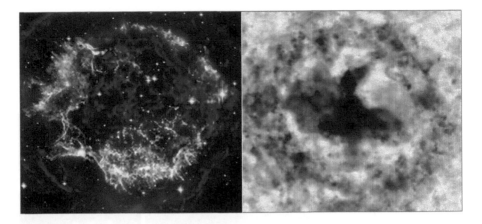

Fig. 8.1. The left panel from a 2007 NASA's Spitzer space telescope detected the infrared signature of silica (sand) in the supernova remnant Cassiopeia A. The light from this exploding star first reached Earth in the 1600s. In the right panel, an example of an exploded neuron forms the Aβ42-immunolabeled (brown) amyloid plaque in the AD brain. The remains of the exploded star and neuron are visible just off-center by the cyan dot and the purple nucleus, respectively. (The left panel credit: NASA/JPL-Caltech/O. Krause (Steward Observatory). The right panel is courtesy of *Curr Pharm Des.* 2006;12(6):677–684.)

disease states that you draw hypotheses, just like pathologists have been routinely doing for decades.

COMPOSITION

The dense-core amyloid plaque contains neuronal debris, but the type of debris present depends on whether that material is sensitive to lysis, another parallel to supernova. As Huggins determined, the spectra of the other planetary bodies he examined shared this remarkable appearance. Apparently that was the pivotal evidence required to make the connection between dying stars and supernova. With regards to the plaques, I am still amazed that in addition to the presence of amyloid in these dense-core plaques, numerous publications have reported finding intracellular neuronal components such as cathepsin D, neurofibrillary tangles, tau, neurofilaments, and ubiquitin. This list has even been extended to show the dispersion of centromeric DNA repeat sequences in the amyloid plaques (Figure 8.2).[13]

If neurons die from the "inside-out," then the therapeutic target is to keep the amyloid from getting inside the neuron. Removing amyloid plaques would be inconsequential since the neurons are already dead.

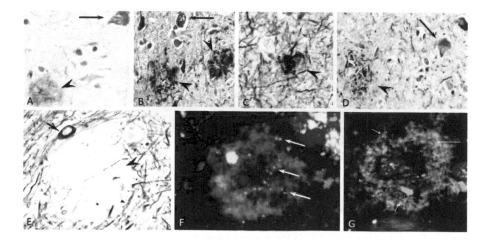

Fig. 8.2. This panel of images in the AD brain shows the detection of neurofibrillary tangles (A), tau (B), neurofilament (C), and ubiquitin (D) in plaques (arrowheads) and in nearby neurons (arrows). Note the absence of MAP-2 immunolabeling in the area of a dense-core amyloid plaque (arrowhead) and the presence of MAP-2 in fibers around the plaque and in a nearby neuron (arrows) in Panel E. In Panels F and G, centromeric DNA repeat sequences are dispersed in areas of amyloid plaques (arrows). (Courtesy of Biotech Histochem. 2010;85(5):133–147.)

If the current amyloid hypothesis is true, then the smaller the plaque, the less likely it would be to have neuronal debris because it did not kill the neuron yet, and, conversely, the larger the plaque it is more likely to have neuronal debris embedded in it. Also, if a plaque did grow to a point to kill the nearby neuron, then one would expect to see neuronal debris to be located to the side of the plaque, or certainly not uniformly distributed throughout the plaque.

For the "inside-out" hypothesis, my data suggest that these dense-core amyloid plaque types have neuronal debris independent of size; remember that the relative size is dependent on (1) the initial size of the neuron as pyramidal neurons vary in size in the brain, (2) potentially the activity of the enzymes released from the neuron, and (3) histological section where grazing sections would appear smaller than midline sections through the spherical plaque (Figures 6.3 and 7.2). I have yet to see extracellular amyloid impinging itself on a neighboring neuron, like a wart on the edge of a neuron (Figures 2.3–2.5). I have never observed neuronal debris on the side of a dense-core plaque type, but only throughout (Figures 2.7, 4.2, and 8.2). In addition, the diffuse amyloid plaque types do not and will never have neuronal debris because they did not form from a dying neuron.

As depicted in Figure 8.1, another parallel is that a remnant star (neutron star) or even a black hole (depending on the star's original size) remains in the middle (core) of the supernova. As I have shown in many images, what remains in the middle area of the dense-core amyloid plaque type is the neuronal nucleus (Figure 2.7) with its intact DNA (Figure 8.2F and G), which actually could be the impetus for the initiation of the inflammatory cascade. Histologically, the opportunity to identify the neuronal nucleus will depend on the sectioning since the dense-core amyloid plaque types are usually spherical (Figure 7.2). An analogy to this would be the slicing of a peach in half to see the pit as compared with the off-center slices of the peach with no pit. The size of the plaque (the bursting/lysis of the neuron) is limited by the brain material and perhaps by the activity of the enzymes released from within the neuron, whereas the size of the supernova (the lysis/explosion of a star) is without impedance in space.

Now that you read this analogy, if you walked through the charred remnants of a burned house, would you still feel that the "carbon" started the fire? Or that the extracellular diffuse amyloid forms a dense-core amyloid plaque or that the extrastellar gases form a star? Is it a coincidence that the remnant star remains in the middle of a supernova as the nuclear debris does in these dense-core plaques? Also, realize that the first supernovae were observed hundreds of years before the connections with stars were appreciated. This journey of connecting cell lysis and dense-core plaques has been 15 years for me so far, and I hope that the discoveries I have made begin to receive serious consideration.

CLOSING COMMENT

The supernova story is a brilliant example of how observation leads to hypothesis, and then scientific fact. The pathway is remarkably similar to the scientific process presented in this book, which unlike that of the supernova example did not begin with a highly endorsed, well-funded hypothesis.

In sum, five points to make concerning supernova:

1. Before spectroscopy, the cosmic vapors of supernova coalesced to become a luminous new star.

2. With the advent of spectroscopy new theories were developed.
3. Star stuff was localized in supernova.
4. The riddle of the nebulae was solved: supernovae are the result of exploding stars.
5. The need to classify nova types was based on their etiology (energy released from a typical nova is less than one-thousandths of the energy of a supernova).

In parallel, five points to make concerning Alzheimer's disease:

1. Before immunohistochemistry (and to date), the extracellular amyloid in the diffuse plaques continues to aggregate to form dense-core plaques.
2. With the advent of immunohistochemistry new theories are developed.
3. Neuronal proteins were localized in dense-core amyloid plaques.
4. The riddle of the amyloid plaque types was solved: diffuse plaques are the result of leaky vessels, and dense-core plaques are the result of exploding neurons; however, to solve the riddle for AD, you have to understand the gravity of too much intracellular $A\beta42$ and how to keep it from getting into the brain and, subsequently, into the neurons.
5. The need to classify plaques is proposed based on their etiology.

REFERENCES

1. http://www.radio.cz/en/section/curraffrs/astronomer-tycho-brahe-was-born-460-years-ago (accessed August 10, 2014).

2. Lankford J. *History of Astronomy: An Encyclopedia.* 1st ed. Routledge; 1996.

3. Hoffleit D. A history of variable star astronomy to 1900 and slightly beyond. *J Am Assoc Variable Star Observers.* 1986;15:77–106.

4. Becker BJ. *Eclecticism, Opportunism, and the Evolution of a New Research Agenda: William and Margaret Huggins and the Origins of Astrophysics* [dissertation]. Baltimore, MD: The Johns Hopkins Unversity; 1993.

5. Fridjung M, Duerbeck H. Models of classical and recurrent novae. In: Hack M, la Dous C, eds. Cataclysmic Variables and Related Objects. Washington, DC: NASA Scientific and Technical Information Branch; 1993:371–412.

6. Maunder EW. Sir W. Huggins and Spectroscopic Astronomy. London/New York: T.C. & E.C. Jack/Dodge Publishing Co; 1913.

7. Hoyle F. The synthesis of the elements from hydrogen. *Monthly Notices R Astronomical Soc.* 1946;106:343–383.

8. Woosley SE. Hoyle & Fowler's nucleosynthesis in supernovae. *Astrophys J.* 1999;525C:924.

9. Sagan C. The Story of Cosmic Evolution, Science and Civilization. Little, Brown Book Group; 1983.

10. Baade W, Zwicky F. On super-novae. *Proc Natl Acad Sci U S A*. 1934;20:254–259.

11. Osterbrock DE. Who really coined the word supernova? Who first predicted neutron stars?. *Bull Am Astronomical Soc*. 1999;33:1330.

12. Murdin P, Murdin L. Supernovae. 2nd ed. Cambridge University Press; 1985:42.

13. D'Andrea MR, Nagele RG. Morphologically distinct types of amyloid plaques point the way to a better understanding of Alzheimer's disease pathogenesis. *Biotech Histochem*. 2010;85(2):133–147.

CHAPTER 9

The Inflammation Cascade

… collateral damage …

Amyloid plaque types are currently named by the morphology of the Aβ42 staining. Functionally, the most important distinguishing feature of plaques is their association with activated astrocytes and reactive microglia. In the event of cell death in the rest of the body, factors are released to alarm the inflammatory cells to open the permeability of the blood vessels to allow more inflammatory cells at the site of the injury to isolate the damage, and then remove the harmful factors. However, in the immune-privileged brain, the inflammatory processes are somewhat autonomous from the body and need to deal with injury on their own. Similarly, cell death triggers the mobilization of the brain's inflammatory microglial cells to ingest the debris, which then activates the astrocytes to help create a glial scar to isolate and heal the area.

As evidenced by the amyloid hypothesis and hundreds of subsequently related papers, microglia are reactive by the extracellular amyloid in the brain, and this has been demonstrated in many publications. However, if this were the sole trigger to cause microglia to migrate to the amyloid plaques, then why wouldn't the microglia be associated with all extracellular amyloid, specifically Aβ42? Why are microglia only associated with certain amyloid plaques? And why wouldn't the positioning of these microglia with these specific plaque types typically be located around the circumference of the plaque? How do these reports explain the presence of microglia in the center of these specific plaque types? And, based on my findings, why are microglia not associated with those "diffuse" plaque types?

At that time, I was invited to debate the subject of microglia's role in the AD brain (Cincinnati, OH, 2003: "Role of microglia in AD does not produce amyloid plaques") and published a review of presentations at this meeting.[1] I presented an "inflammatory cascade" to try to explain primary and secondary causes of neuronal death in the AD brain, which

Bursting Neurons and Fading Memories. http://dx.doi.org/10.1016/B978-0-12-801979-5.00009-6

still confuses the AD pathology. Although two-dimensional, static histological images have their limits and are suspect to overinterpretation, I'd like you to consider observing any image of the night sky. Any two-dimensional image of the night sky presents a timecourse of new and old, of stars near and far. The brain also contains a library of such events with healthy and unhealthy cells, but unlike the time-lapse images of the night sky, to compare brain images, you compare AD with non-AD, normal control brain tissues.

One can presume that the diffuse amyloid plaques are inconsequential because they are occasionally detected in age-matched, nondemented control brains, and they are not associated with inflammatory cells. Removing this diffuse, benign, extracellular amyloid between the neurons in the brain will not change cognitive function since it did not form from dying neurons. Conversely, it is much easier to realize that memory loss and other cognitive changes would be more associated with neuronal death via necrosis that triggers inflammation.

Based on my observations, gliosis seems to be triggered by the lytic death of the neurons. Once the cell dies, it incidentally releases its contents, some of which activate the microglia to mobilize to the area to ingest or phagocytize the cellular debris. While the microglia are present, they then release factors that activate the local astrocytes to extend their processes to create a scar that appears like a web as it tries to fill in the hole left from the dead neuron. Interesting to note is that those same local astrocytes then release factors to deactivate the microglia (Figure 9.1).[2]

Unfortunate to this process is that those released factors, which may be specific to the astrocytes or microglia, harm neighboring neurons causing them to die as collateral damage from the processes of inflammation. Imagine the impact on neighboring neurons whose processes once ran through the area that is now a toxic area of neuronal debris and gliosis. Three-dimensionally speaking, perhaps hundreds of neurons are negatively impacted; some can tolerate the loss of a minor process, while others are so impacted that they also die. Again, envision the MAP-2-negative area, or even the evidence presented in the H&E slide (Figure 2.1), like a hole in a sponge, or a fire causing ancillary damage making it even more difficult to trace the initial pathological event.

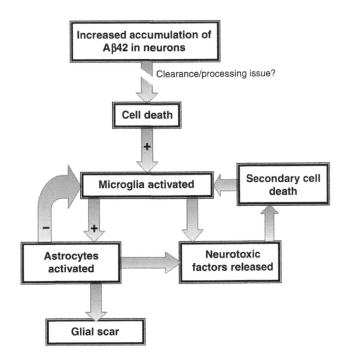

Fig. 9.1. *A proposed pathway of inflammation in the AD brain showing inhibitory (−) and stimulatory (+)*
pathways of primary and secondary cell death consequences. Based on substantial evidence, it was proposed that
one of the primary activators of microglial activations originates from dying or lysed cells. Subsequent microglial
activation can trigger astrocyte activation (+) (although astrocytic activation may occur independent of microglial
activation), which in turn can inhibit microglial activation (−). However, both reactive microglia and astrocytes
secrete factors that are toxic to neurons, thereby contributing to a pathological cascade. (Courtesy of *Neurobiol*
Aging. 2004;25:675–683.)

Some neurons are dying by necrosis directly from engorging amyloid resulting in the pathological, dense-core, MAP-2-negative amyloid plaque, while others are dying by necrosis directly from the toxic debris from the lysed neuron. Still others are dying unobtrusively by apoptosis either indirectly from not being able to recover from the loss of some of their processes in the wake of the neuronal bursting event or by other unknown events.

REFERENCES

1. D'Andrea MR, Cole GM, Ard MD. The microglial phagocytic role with specific plaque types in the Alzheimer disease brain. *Neurobiol Aging.* 2004;25(5):675–683.

2. D'Andrea MR, Reiser PA, Gumula NA, Hertzog BM, Andrade-Gordon P. Application of triple-label immunohistochemistry to characterize inflammation in Alzheimer's disease brains. *Biotech Histochem.* 2001;76(2):97–106.

CHAPTER 10

Innocent Aβ42

… present throughout life …

Sometimes you just have to trust your data and charge ahead, even in the face of dogma. How can Aβ42 be present in "normal" neurons especially since the Aβ42 form of amyloid is understood by the amyloid hypothesis to be the pathological form of amyloid that causes neuronal death leading to AD (Figure 10.1)?[1] The simple detection of Aβ42 in normal neurons is inconsistent with the current understanding that it is toxic, unless perhaps, there is too much of it.

It has been a challenge not only to prove the presence of fatal levels in Aβ42 in the AD neurons but to prove beyond reasonable doubt that Aβ42 is also present in normal neurons. This somewhat surprising discovery inferred that Aβ42 is a normal protein. I said to myself, how can I convince researchers to understand that Aβ42 is not toxic while still playing a pathological role in AD? The detection of Aβ42 in normal tissues[1,2] suggests that the pathological role of Aβ42 is its unregulated presence in the AD brain. It's all about data, and methods; can you reproduce your findings over and over again? Can others reproduce your findings?

If you search for amyloid function, you will see conflicting information. Some papers report that its primary function is unknown, while others report its role as a modulator of synaptic plasticity through wiring and rewiring of the brain. Specifically Aβ is present in the brain of symptom-free subjects suggesting an important physiological role and synaptic activity directly produces the release of Aβ at the synapse.[3] However, the life cycle of Aβ42 begins with APP, the amyloid precursor protein that is also present in many tissues. The processing of APP by the cells ultimately yields smaller fragments that are named based on their amino acid sizes, such as Aβ42, Aβ40, and so forth. And as mentioned earlier, it is the Aβ42 species that is highly fibrillogenic and is deposited in the AD brain. Nonetheless, Aβ42 was clearly present in normal neurons (see Figure 10.1), and smooth muscle cells that cover the blood vessels, specifically arteries (Figure 10.2).

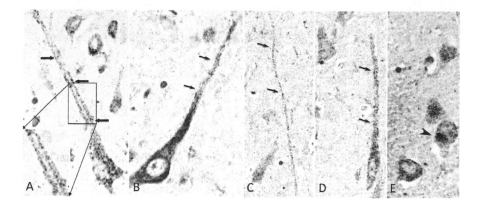

Fig. 10.1. Representative images of Aβ42 immunolabeling in cortical sections from age-matched, nondemented controls. Note the Aβ42 distribution within the cell body and dendritic processes (arrows) and within various nearby neurons as well as the tendency for neurons to sequester their intracellular Aβ42 away from the entrance into axons and dendrites (arrowhead). (Same figure as Figure 2.5 is reprinted in this chapter for convenience.) (Courtesy of *Biotech Histochem.* 2010;85(5):133–147.)

Based on my reported data, Aβ42 is present throughout life. The presence of Aβ42 in normal cells suggests that Aβ42 is not toxic and may actually play a normal physiological role in synaptic plasticity. However, the amount of Aβ42 in the protected area in the brain is another fundamental circumstance that also requires some explaining.

To date, the presence of Aβ42 in the brain is thought to be a product of the neurons. It is believed that the neurons make it, and secrete

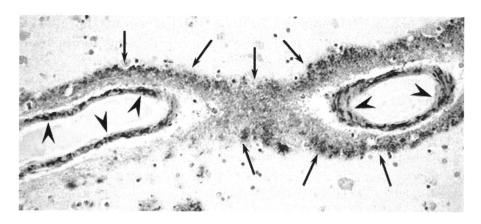

Fig. 10.2. Presence of Aβ42 in vascular smooth muscle cells (arrowheads) and appears to leak away from the vessels to form a diffuse cloud of extracellular Aβ42 (arrows) in the AD brain, also known as the diffuse amyloid plaque. Further described in Biotech Histochem. *2010;85(5):133–147.*

it where it deposits outside between the neurons and grows becoming more fibrillar and toxic to eventually kill neighboring neurons. The de novo synthesis (made from the beginning) of amyloid in the neurons of the brain is part of the current thinking, and therefore no real attention is given to the other possibility that perhaps the amyloid, specifically Aβ42, is originating from outside the brain through the cardiovascular system. I am not stating that neurons do not make amyloid or Aβ42, but perhaps they do it at the rate required for normal physiological function as in synaptic plasticity.

So, why innocent? It's the BBB's fault! As noted, Aβ42 is a normally functioning physiological protein that just happens to be highly sought by receptors on neurons in the brain, and the BBB normally functions to regulate the entry of material into the brain. However, Aβ42 is not totally innocent either. As per the house fire analogy, Aβ42 happens to be left at the place of the crime (aka carbon residue) but is unfortunately framed. So, it is indirectly guilty and because of a dysfunctional, more porous BBB, too much of the Aβ42 enters the AD brain without regulation leading to the unexpected demise of the neurons.

REFERENCES

1. D'Andrea MR, Nagele RG. Morphologically distinct types of amyloid plaques point the way to a better understanding of Alzheimer's disease pathogenesis. *Biotech Histochem.* 2010;85(2):133–147.

2. Wang HY, Lee DHS, D'Andrea MR, Peterson PA, Shank R, Reitz A. β-Amyloid1-42 binds to α7 nicotinic acetylcholine receptor with high affinity: implications for Alzheimer's disease pathology. *J Biol Chem.* 2000;275(8):5626–5632.

3. Parihar MS, Brewer GJ. Amyloid-β as a modulator of synaptic plasticity. *J Alzheimers Dis.* 2010;22:741–763.

CHAPTER 11

The Alpha 7 Nicotinic Acetylcholine Receptor

… not equipped to regulate …

Aβ42 can be detected in normal and AD neurons. In AD, the neurons fatally and uncontrollably accumulate this material to the extent of engorging with Aβ42. It is as if they had never been exposed to unregulated supplies entering the brain. Slowly, perhaps over many years, there comes a point when there is so much of this material in the neuron that it cannot possibly function, and dies. While I do not know why, I remember reading years ago that thousands of mitochondria in a cell can be trafficked from areas within the cell to another area within one second. Now imagine trying to maintain normal order or homeostasis while having most of your living space occupied by junk to the point where you cannot even see the counters, the furniture, and the doors. Ultimately, the cell cannot function and it dies perhaps leaving the cell maimed, although it's not clear to me how the cell specifically dies, be it loss of the ATP pumps leading to loss of internal pressure, or some similar loss of critical function. As the cell dies, the cell membrane cannot maintain its barrier and breaks apart or bursts leaving the intracellular components exposed and now loose between the cells of the brain. Then, the once harmless encapsulated enzymes that normally digest proteolytically sensitive proteins inside the cell begin slowly digesting the brain for a finite time and distance as per tight range of dense-core plaque sizes. This area of neuronal carnage typically includes a nucleus leaving the remaining indigestible neuronal debris such as amyloid, tau, neurofilaments, ubiquitin, and cathepsin D, in the space previously occupied by the mighty neuron.

So, how is the Aβ42 getting into the neuron? And why are some neurons resistant to such fate of death by Aβ42?

There are many types of neurons in the brain, in the same way there are cell types in the body. For the most part, neurons are grouped by their functions that align with the types of transmitters they possess. For example, some neurons operate through the acetylcholine pathway, while others through adrenergic pathways. The particular type of

Bursting Neurons and Fading Memories. http://dx.doi.org/10.1016/B978-0-12-801979-5.00011-4

neurons that appear to be vulnerable to Aβ42 have a particular receptor on their cell membrane called the alpha 7 nicotinic acetylcholine (α7) receptor, as discussed earlier. This postsynaptic membrane receptor is a neuronal homopentameric cation channel that is highly permeable to Ca^{2+}, widely distributed in the nervous system, highly expressed in basal forebrain cholinergic neurons that project to the hippocampus and cortex, and involved in cognition and memory. It was this specific project that brought me into the AD field as I was tasked to understand the spatial relationship between the distributions of the α7 receptor and Aβ42 in the AD brain. I demonstrated the distribution of the α7 receptor in the AD tissues and was able to show colocalization of the α7 receptor with Aβ42 in plaques and in some neurons (Figure 2.3).

This research had demonstrated the very high binding affinities of the α7 receptor and Aβ42, suggesting that if Aβ42 is present, it will bind to the α7 receptor.[1] Furthermore, this extraordinary stability of the α7 receptor:Aβ42 complex may explain the observed accumulation of the two proteins in the cell body of the AD neurons, and may also be related to its resistance to normal proteolytic clearance.

These results were reinforced by using cells that were transfected with the α7 receptor while in culture dishes in vitro, and when Aβ42 was added to the supernatant in the dishes, those cells internalized the Aβ42 that was inhibited using α-bungarotoxin, a specific α7 receptor antagonist (Figure 11.1).[1] A decreased amount of α7 receptor in sporadic AD brain tissues correlated with the loss of neurons.[2] Hence, in addition to Aβ42, the α7 receptor also plays an important role in the pathology of AD.

This observation was further pursued with the same α7 receptor transfected cells but this time, the fluorescently labeled Aβ42 was added to the cells in vitro and over time, they began to die. Similarly, if an inhibitor of the α7 receptor was added to the culture dishes, the internalization of Aβ42 was blocked and the cells lived.[3] Additionally, if an inhibitor of endocytosis was added to the cells in the presence of Aβ42, the internalization of Aβ42 was also blocked.[3]

These studies suggested that extracellular Aβ42 will bind to the α7 receptor on specific neurons and the cells will internalize the Aβ42 by the processes of endocytosis as evidence by the colocalization of α7 and Aβ42 in AD neurons (Figure 2.3). At that time in 2006, I was invited to

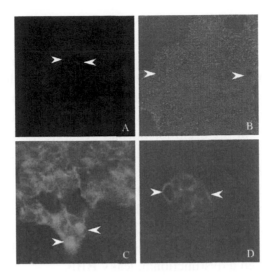

Fig. 11.1.   Fluorescent Aβ42 binds to α7 receptor. Transfected α7SK-N-MC cells were grown on 22-mm² plastic coverslips. Fluorescent Aβ42 was added to the transfected cells and incubated in the presence or absence of 1 mM α-BTX. Fluorescent Aβ42 added to wild-type cells was used as a control (data not shown). The slides were mounted with an antifade 4,6-diamidino-2-phenylindole fluorescent medium and examined under a Zeiss confocal laser scanning fluorescent microscope. Texas Red staining for α7 receptor (arrowheads) is shown in wild-type SK-N-MC cells (A) and α7SK-N-MC cells (B) confirming successful transfection. In Panel C, fluorescent Aβ42 (green) is detected on α7SK-N-MC (arrowheads). In Panel D, reduced fluorescent Aβ42 (green) detection in the presence of 1 mM of α-BTX. Similar results were obtained in six separate experiments. BTX, bungarotoxin. (Courtesy of J Biol Chem. 2000;275(8):5626–5632.)

submit a review article on this pathway and suggested targeting the α7 receptor to reduce amyloid accumulation in AD neurons.[4]

The presence of the α7 receptor on smooth muscle cells and specifically on vessels in the brain provided further proof of the presence of Aβ42 in the smooth muscle cells of the vessels where, once again, they form vascular plaques.

When considering this new hypothesis, you might wonder of the real purpose of Aβ42 especially since there is so much of this material in the brain. Based on the binding affinity of Aβ42 to the α7 receptor and its normal physiological function, one could propose that Aβ42 is typically sparse and regulated in the brain and that even a miniscule amount would be enough to trigger activity for normal synaptic plasticity and formation like learning. As an unwritten law, I would presume that the binding affinity is directly inverse to the abundance of the ligand, meaning the less abundant the ligand, the higher the affinity of the receptor. The

AD neurons with less internalized Aβ42 may represent the early stages of its "unexpected" trickling into the brain that may initially cripple the neurons thereby losing the ability to sustain their far-reaching processes leading to synaptic collapse, thought to represent the early stages of AD. Over time, unregulated internalization of excessive amounts of Aβ42 eventually leads to cell death. In other words, there may be a delicate balance of Aβ42 in the brain for normal synaptic modeling, and once this balance is shifted by uncontrolled outside contributions of Aβ42 into the brain, the neurons are incapable of regulating their internalization of Aβ42. The once physiologically important Aβ42 to regulate synaptic plasticity becomes inadvertently toxic not due to its fibular nature, but due to its excessive presence in the brain and high affinity to the α7 receptor, and the only possible source would have to be the blood supply specifically through a dysfunctional, leaky BBB.

In sum, it appears that the α7 receptor-positive neurons were not equipped to regulate internalizing the unnaturally excessive levels of Aβ42 in the brain that eventually led to their demise. Therefore, inhibiting the internalization has to be one of the most important therapeutic areas of focus for AD research: if you stop it from going into the neurons, you are protecting vulnerable neurons from inevitable death thereby reducing secondary cell death due to the local lytic event and through the subsequent processes of inflammation.

REFERENCES

1. Wang HY, Lee DHS, D'Andrea MR, Peterson PA, Shank R, Reitz A. β-Amyloid1-42 binds to α7 nicotinic acetylcholine receptor with high affinity: implications for Alzheimer's disease pathology. *J Biol Chem.* 2000;275(8):5626–5632.

2. Wang H-Y, D'Andrea MR, Plata-Salaman CR, Lee DHS. Decreased α7 nicotinic acetylcholine receptor proteins in sporadic Alzheimer's disease brains. *Alzheimers Lett.* 2000;3(4):215–218.

3. Nagele RG, D'Andrea MR, Wang H-Y. Intraneuronal accumulation of β-amyloid 1-42 is mediated by the α7 nicotinic acetylcholine receptor in Alzheimer's disease. *Neuroscience.* 2002;110(2):199–211.

4. D'Andrea MR, Nagele RG. Targeting the alpha 7 nicotinic acetylcholine receptor to reduce amyloid accumulation in Alzheimer's disease pyramidal neurons. *Curr Pharm Des.* 2006;12(6): 677–684.

Immunoglobulin: Another Perpetrator

... the single, unifying connection is the dysfunctional BBB ...

Is AD an autoimmune disease? It certainly plays a role. I did not come to this unconventional belief until I was trying to prove that amyloid, specifically Aβ42, gains entry into the brain through a dysfunctional BBB, and that this amyloid pools in focal areas of the brain, thereby explaining the origin of the benign or diffuse amyloid plaque type. I considered a new approach to prove this hypothesis by using a marker that is typically found in the blood stream and scarcely reported in the brain. Hence, if this particular marker is also in the brain, then it could further support the notion that first, there is a dysfunctional BBB, and second, vascular elements would enter the brain without regulation.

As you may know, the normally functioning BBB isolates and protects the brain from the rest of the body, which is typically a pharmaceutical problem: because the barrier is so tight and selective, it is a challenge for medicines to gain entry to their CNS targets. Consequently, the brain cannot rely on the body's inflammation system for help and therefore has developed its own police system to deal with inflammation. The microglia inflammatory cells of the brain constantly scavenge the brain for cell debris and infectious agents, while the primary function of other supportive cells, the astrocytes, is to maintain the BBB while repairing injured areas by extending their processes that eventually form a glial scar. Since circulating antibodies do not enter the brain through the barrier, microglia must be able to quickly recognize foreign substances and engulf them.

Of all the possible vascular markers, I chose a generic immunoglobulin (Ig) marker to detect all Ig classes (IgA, IgG, IgM), commonly referred to as pan-Ig.[1] Since Igs are too large to cross the BBB, if I could detect Igs in the AD brains, it could further support the possibility that amyloid could also enter the brain through the BBB if it was dysfunctional. Typically, Igs are found only in the vessels of the circulatory system and at sites of inflammation in tissues, but never, or rarely, in the

Bursting Neurons and Fading Memories. http://dx.doi.org/10.1016/B978-0-12-801979-5.00012-6

brain because of the functioning BBB. The mere presence of any Ig in the brain would be akin to a large truck entering through the front door of your house. The barrier does allow some of the vascular components to enter the brain for nourishment, but those components are restricted based on characteristics of size, hydrophobicity, and charge. However, if the barrier is dysfunctional, these restrictions are lifted and anything could enter independent of size or charge. Therefore, I designed a new set of IHC assays to detect Ig in the AD brain tissues, and I would also assay normal, nondemented, age-matched control brains.

I remember first reading these slides and to my satisfaction, Igs were detected in the AD brain as diffuse clouds of immunolabeling outside the vessels (Figure 12.1).[1]

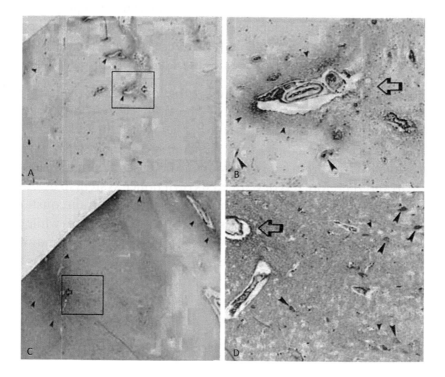

Fig. 12.1. Representative images include Ig immunolabeling in "control" (A, B) and AD (C, D) cortical tissues at low and high magnifications. Areas in the black boxes in the lower-magnification micrographs (A, C) indicate areas of higher magnifications (B, D). Arrowheads show areas of Ig immunoreactivity in the brain (A, C), which appears to clearly originate from vessels in the "control" tissues (A, B). At higher magnification of the "control" brain (B), small arrowheads show areas of Ig immunolabeling around cerebral vessels and large arrowheads show areas of Ig in nearby vessels that do not show percolating Ig immunoreactivity in the neighboring brain parenchyma. Large arrowheads in Panel D identify Ig-positive neurons in the AD tissue among Ig-negative neurons (small arrowheads). (Courtesy of Brain Res. 2003;982:19–30.)

I quickly viewed the positive and negative controls, the preabsorption antibody controls, and all appeared validated.[1] These data were quite compelling and could provide the morphological evidence to not only further support previous AD cardiovascular factors such as hypertension, stroke, and diabetes but also provide a mechanism to the origin of the excessive and unregulated amounts of Aβ42 in the AD brain while defining the origin of those diffuse plaques that were clearly associated with vessels.

I then studied several control brain tissues and noticed smaller areas of amyloid in the brain parenchyma (Figure 12.1), but it was quite clear that vascular Igs were more prominent in the AD brain parenchyma in comparison to age-matched, nondemented control brain tissues. I certainly did not expect to find Ig in the normal brain tissues, but perhaps over time, our BBBs become leaky, underlining the AD cardiovascular disease risk factors but perhaps there must be additional risk factors. Nonetheless, there must be additional and exacerbating risk factors that turn the ultimate shift from the brains of normal to AD.

To quantitate the Ig immunolabeling, I used computer-assisted image analysis to measure the amounts of Ig present in all of the brain tissues and determined that although some Ig labeling was detected in the control brain tissues (about 8% of the brain area), significantly more was detected in the AD brain, about 55% of the brain area.

IMMUNOGLOBULIN NEURONS

This observation alone does not suggest that AD is an autoimmune disease, but when I started to examine the AD tissues at higher magnifications under the microscope, I observed something I had not predicted. I noticed Ig-positive and Ig-negative immunolabeled neurons side by side. The fact that they were side by side suggested this was no artifact, or a false-positive or false-negative effect, but that some neurons had these vascular-derived Igs inside while other neurons did not (Figure 12.2).

As your eyes look at this figure, I'm sure it's the darkly labeled neurons that capture your attention, but for me, it is that ghostly, Ig-negative neuron. This Ig-negatively labeled neuron is the key to the suggestion that AD may be an autoimmune disease. Among a sea of Ig labeling in the brain and next to two Ig-positive neurons, it is the absence of Ig labeling in this neuron that actually serves as internal negative control

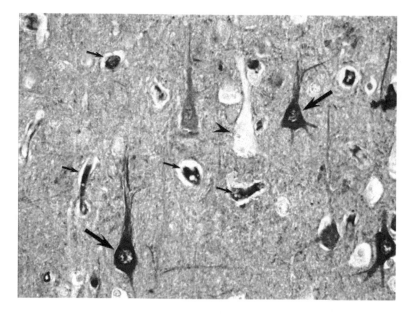

Fig. 12.2. This image (one of my first) shows the presence of Ig-positive neurons (large arrows) in an area of extracellular brown, Ig immunolabeling in the AD brain. Note the presence of a single Ig-negative neuron (arrowhead). Small arrows show presence of vascular Ig immunolabeling in vessels.

for the technical conditions thereby ruling out potential technical artifacts. In other words, not everything was nondiscriminately positive.

I noticed that these Ig-positive neurons were predominantly located in areas of Ig labeling in the brain, which made perfect sense. It was logical to believe that the Igs leak out of the blood vessels into the brain parenchyma due to a dysfunctional barrier, but why would some neurons internalize to become Ig-positive, while other nearby neurons, bathed in the same Igs in the brain, are without labeling? Perhaps the Igs are binding to specific targets not shared by all neurons suggesting that the Igs recognize specific features on a neuron by type.

After hours of counting the Ig-positive and Ig-negative neurons from one brain to the next, and then comparing those numbers with the data obtained by determining the percentage of Ig labeling in brain parenchyma for each tissue in control and AD brains, there was a positive correlation stating that the number of Ig-positive neurons increased as the percent of Ig labeling in the brain parenchyma increased, which again makes perfect sense. Also, the percentages of Ig-positive neurons were significantly higher ($P < 0.001$) in the AD brain (entorhinal cortex and hippocampus) as compared with those in the control brain.[1]

DEGENERATING Ig-POSITIVE NEURONS

The next key observation was that many of those Ig-positive neurons were degenerating.[1] How can you tell if a neuron is sickly or not by examining tissue on a microscopic slide? It was based on morphological features of pending cell death that include atrophy or withering neuronal processes, and dense, pyknotic nuclear chromatins that have dark, solid purple nuclei in some of the Ig-positive neurons. These sickly neurons were in stark contrast to neighboring Ig-negative neurons with normal neuronal features, and nuclei with euchromatic chromatin and prominent nucleoli that suggested normal nuclear transcriptional activity (Figure 12.3).

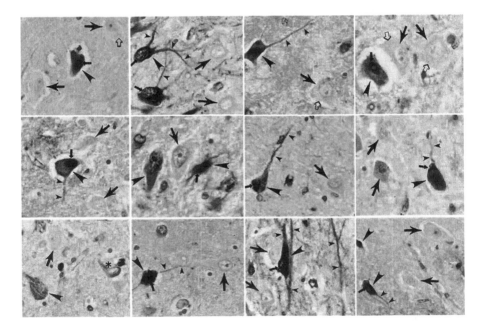

Fig. 12.3. Ig-positive neurons (large arrowheads) and Ig-negative neurons (large solid arrows) are observed in "control" (top left figure) and AD (all other figures) brain tissues. Ig immunolabeling is observed diffusely in the perikaryon (large arrowheads) and nucleoplasm (small solid arrows) of most affected neurons. It is interesting to note the presence of morphologically normal Ig-negative neurons, with their prominent nucleolus, euchromatic nuclear chromatin, and well-defined cell perikaryon (large solid arrows). The Ig-positive neurons show degenerating apoptotic features such as cell atrophy (large arrowheads) and degenerating processes (small arrowheads), condensed, pyknotic nuclear chromatin to the point at which the normal nuclear appearance is not apparent (small solid arrows). Open arrows show the presence of unstained cellular lipofuscin. An asterisk in the lower left panel shows Ig immunoreactivity in the shape of a neurofibrillary tangle (neuronal cast). (Courtesy of Brain Res. 2003;982:19–30.)

Whether the neurons were dying due to the Ig or by another mechanism is not clear, but Ig seemed to be a marker of these dying neurons. So what comes first? Is a normal neuron Ig-negative first, and then when it becomes Ig-positive, degenerates and dies? Or was the neuron already dying and then became Ig-positive as an artifact? Or were those dying neurons simply an artifact due to postmortem manipulation or trauma in brain tissue?[2] And what is the relationship between Ig-positive neurons and Aβ42?

And so the next step was to determine if there were any Ig-negative neurons that appeared sickly; if so, then it would be fair to propose that the Ig labeling is independent of the neuron's health. I already noted that Ig-positive neurons are degenerating, but if I was able to observe morphologically healthy Ig-positive neurons, then it could suggest that these neurons will die over time, and that the conversion of a neuron from Ig-negative to Ig-positive could be related to cell death.

Once again, back to the microscope and to reread all the slides to accumulate another session of countless microscope hours. After reviewing over 900 neurons (18 ADs and 13 controls), less than 1% (2 of 900) of the Ig-negative neurons displayed neurodegenerative features.[1] Furthermore, an average of 40% on the Ig-positive neurons showed these morphological signs of degeneration. These data told me that those Ig-positive neurons will eventually die and Ig detection in a neuron appears to be a marker of pending cell death (Figure 12.4).

Unlike those Aβ42-overburdened neurons previously described to be dying from cell lysis forming the dense-core plaque, many of these Ig-positive dying neurons did not appear to be lysing or associated with gliosis suggesting that they may be dying by apoptosis and not due to necrosis as described in the formation of those dense-core, amyloid plaques.

This new pathological process adds another bit of complication to the AD pathology:

- Neurons dying by cell lysis due to overaccumulation of Aβ42
- Neurons dying in the wake of the cell lysis
- Neurons dying as a consequence of inflammation

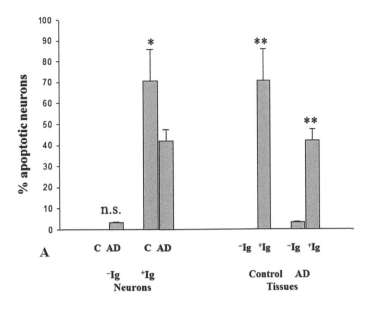

*Fig. 12.4. Bar graph shows significant increases (P < 0.001) in the percentage of Ig-positive neurons with apoptotic morphological features as compared with the Ig-negative neurons (*n = 10 AD *and* n = 4 *age-matched control brain tissues).* (Courtesy of *Brain Res.* 2003;982:19–30.)

- And now some neurons dying possibly by apoptosis from the autoimmune, Ig-positive, detection, which in itself would also trigger an inflammatory response by the microglia

There may not be a single underlying pathology in AD, but it does seem that the single, unifying connection in these neuronal death scenarios is the dysfunctional BBB.

I needed to confirm if those neurons are indeed dying through the processes of apoptosis, and the way to do this was to see if they are expressing apoptosis-related proteins such as activated caspase 3, an enzyme reported to be present only in apoptotic cells. So I performed double IHCs to detect Ig and activated caspase 3 and, as suspected, many of those Ig-positive neurons also expressed activated caspase 3, while none of the Ig-negative neurons expressed activated caspase 3.[1]

This possibility of AD neurons dying by apoptosis was previously reported, but now I was able to confirm and introduce the possibility that they may be dying through specific antineuronal Igs in an autoimmune process. How else can you explain the presence of Igs in some neurons, among many Ig-negative neurons?

I also noticed the presence of Ig in astrocytes that were generally located around those "leaky" vessels bathed in extracellular Ig. Astrocytes have several functions in the brain. Of their various functions, most importantly they support the vascular endothelial cells that form the BBB and have a role in the repair and scarring processes of the brain following injury such as neuronal death. Interestingly, these Ig-positive astrocytes did not show any signs of activation or gliosis, degeneration, or caspase 3 detection.[1] The Ig was localized in the cytoplasm of the astrocyte suggesting that the astrocytes internalized the extracellular Ig through endocytosis, a well-known normal astrocytic function. These observations also suggest that the Igs do not recognize astrocytes as they do for specific neuronal types.

I found all of this information serendipitously while trying to prove that the BBB is leaky as I decided to detect Igs to make the case. It could have been any number of vascular components such as fibrin, platelets, and so forth, but because it was Ig, I was able to observe all of that information above. I submitted the paper to *Brain Research*, who accepted the paper with very few comments. As of early 2014, this paper has been cited in only about 60 research papers, but I feel in time, this number will increase as more and more AD researchers gain a wider acceptance.[1]

I was still puzzled to understand the connection between the formation of the amyloid plaques through neuronal lysis and Ig-positive neurons. I already believed that amyloid leaks out of the dysfunctional BBB and begins to accumulate in the specific neurons via the $\alpha 7$ receptor over time to the point that the neuron becomes so engorged that it lyses and dies thereby activating gliosis, but how do the Ig-positive neurons play in this process? Is there a connection between Ig-positive neurons and $A\beta 42$-overburdened neurons? Are they one and the same?

REFERENCES

1. D'Andrea MR. Evidence linking autoimmunity to neuronal cell death in Alzheimer's disease. *Brain Res.* 2003;982(1):19–30.

2. Jortner BS. The return of the dark neurons. A histological artifact complicating contemporary neurotoxicologic evaluation. *Neurotoxicology.* 2006;27(4):628–634.

CHAPTER 13

Add AD to the List of Autoimmune Diseases

… resembled that of another war zone …

I wanted to present this theory to a journal solely based on medical hypotheses. I posed the title "Add AD to the List of Autoimmune Diseases" in the hope to provoke intrigue and excitement in the AD field while drawing attention to the *Brain Research* paper.[1] Although this paper has not been cited often, I strongly believe that over time, this number will increase as the field considers this additional hypothesis. Autoimmunity is not a novel concept in human diseases. In fact, several diseases are well-documented autoimmune diseases of the central nervous system. For those without much background in this area, as the name implies, it is immunity against itself. For various unclear reasons, the body can develop antibodies to attack normal targets, not just those foreign to the body as in typical infections. As cells die, by either injury or inflammation, cellular debris circulates the body, which elicits a response since the body's immune system has not typically seen the inner cellular components. As such, the body sees such "inner parts" as foreign and mounts an attack on this target. Unfortunately, some of these cellular antigens are normal and the next time the body's surveillance system comes into contact with these types of antigens, an attack is mounted. This is the case of many diseases, including Grave's disease, systemic lupus erythematosus, rheumatoid arthritis, scleroderma, and others more specific to the brain such as Huntington's chorea, multiple sclerosis, Sydenham's chorea, and cerebral lupus.[1,2]

As another example, Rasmussen's encephalitis (RE), an autoimmune progressive childhood disease, is characterized by severe epileptic seizures, hemiplegia, dementia, and inflammation of the brain with progressive destruction of a single cerebral hemisphere due to the undesired presence of autoantibodies to the glutamate receptor 3. Interestingly, this is reportedly due to the breakdown of the BBB; however, it was also hypothesized that seizure discharges themselves may damage the BBB to allow those antibodies into the brain to promote the development of RE.[3,4] Similar to the Ig-positive AD neurons, the Ig labeling in RE brain

Bursting Neurons and Fading Memories. http://dx.doi.org/10.1016/B978-0-12-801979-5.00013-8

tissues was also observed in neurons and in their processes in association with complement membrane attack complex immunoreactivity leading to neuronal damage, also reported in myasthenia gravis.

Not much longer after I published this hypothesis, the *BMJ*[5] commented that:

> *The failure to find a sole pathological event that triggers AD hasn't deterred some from believing there's still an undiscovered single event that starts the whole process. The recent demonstration of immunoglobulin and complement in brain tissue from people with Alzheimer's disease raises the possibility that the presence of anti-neuronal auto-antibodies found in serum (previously thought to be of no significance) points to it being an autoimmune disease. One possibility is that a critical dysfunction of the blood–brain barrier allows these auto-antibodies to access and kill their desired target cells.*

After reading the literature about RE and the association with complement, a system of molecules that work together to help dissolve and remove foreign cells, I wondered if the Ig-positive neurons were also dying through classical antibody-dependent complement pathways. Although these processes are better explained in textbooks, basically once an antibody binds to its target, there are several other steps that occur before the cell dies. Some of these steps require the orchestration of various components of the classical pathway that include complement products such as C1q, a specific component of this antibody-induced, complement pathway, and C5b-9, which is a marker of the terminal step in the complement pathway that represents the membrane attack complex ultimately responsible for the death of the cell. These complement products were also detected on neurons in RE and myasthenia gravis.

Based on those same IHC methods and AD brain tissues, I was able to verify that C1q and C5b-9 were also detected only in Ig-positive neurons providing evidence for the presence of the classical antibody-dependent complement pathway of cell death to the Ig-positive neurons.[6]

I was very curious to understand the spatial relationship of the microglia to these Ig-positive degenerating neurons because if these Ig-positive neurons are dying, perhaps the microglia are associated with these neurons? Using double IHC to detect microglia (red stained) and Ig (brown stained), representative images in Figure 13.1 show that reactive microglia were consistently observed within very close proximity

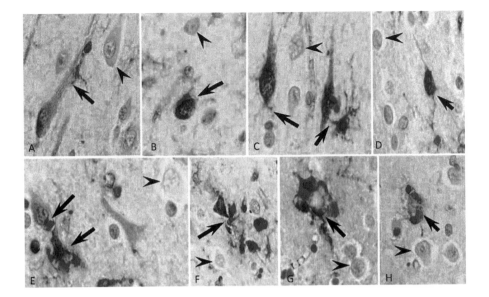

Fig. 13.1. Representative double IHC images show the presence of red-labeled, HLA-DR-positive, reactive microglia in contact or in association with brown-labeled, Ig-positive neurons (arrows) in the AD entorhinal cortex. All nuclei are stained purple from the hematoxylin dye. Arrowheads show the presence of nearby Ig-negative neurons without associated reactive microglia. (Courtesy of Am J Alzheimers Dis Other Demen. 2005;20(3):144–150.)

(Figure 13.2) to many Ig-positive neurons, which were not readily observed near surrounding Ig-negative neurons.

Although again subject to overinterpretation, the microglia detected in close association to the Ig-positive neurons may represent an early stage of identifying cells for impending death. For example, the distinctive signal of neuronal apoptosis is the release of active form of matrix metalloproteinase-3 that activates microglia and subsequently exacerbates neuronal degeneration.[7] Furthermore, phagocytes, such as microglia, efficiently remove apoptotic cells by the recognition of surface receptors such as scavenger receptor A (CD36) and the phosphatidylserine receptor (CD68) as well as components of the complement system.[8] The arrangement of the microglia and Ig-positive neurons once again resembled that of another war zone, where several red-labeled, reactive microglia appear dramatically engaged with the degenerative Ig-positive neurons among the lack of associated reactive microglia with the Ig-negative neurons. Even when I carefully used image analysis to measure the distances between the microglia processes to Ig-positive

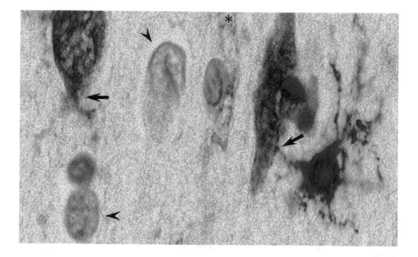

Fig. 13.2. A digital enlargement of Panel C in Figure 13.1 clearly shows the close association of the red-labeled microglia processes with the brown-labeled Ig-positive neurons (arrows). Only some of the red-labeled microglial processes are present in this two-dimensional section for the left area, and note the lack of red-labeled microglia processes with the nearby Ig-negative neurons (arrowheads). The asterisk shows the presence of residual brown-labeled Ig in a nearby capillary. Personally, this is one of my all-time favorite AD images in this book. Besides the presence of the two internal negative-control neurons (arrowheads), the right half of this image is striking and shows the "reach out and touch" of the microglia to the affected neuron and poetically reminds me of the Michelangelo's finger of God. (Courtesy of *Am J Alzheimers Dis Other Demen.* 2005;20(3):144–150.)

and Ig-negative neuronal nuclei, reactive microglia were significantly ($P < 0.001$) more associated with Ig-positive neurons than Ig-negative neurons. In support, complement produced locally by reactive microglia was reactive on the membranes of neurons in Huntington's disease, contributing to neuronal death as well as proinflammatory activities.

The data in this report suggest that populations of Ig-positive neurons are dying via the classical complement pathway and can be employed in the diagnosis and treatment of AD,[9] and that microglia are preferentially associated with many of these degenerating neurons possibly as an attempt to remove these affected dying neurons. The presence of neuronal specific antibodies alone would not appear harmful unless there is a BBB breach, which again initiates another inflammatory cascade that will undoubtedly inflict additional neuronal cell death independent of immunoglobulin and complement.

Soon after this third publication concerning AD and autoimmunity, I was determined to see if it was possible to identify the target of this autoantibody or autoantibodies because the presence of this auto-antibody, once characterized, should provide a new therapeutic target to treat and possibly prevent AD independent of amyloid. It was clear that the Ig recognizes and binds to specific neurons, and then after binding, it was internalized into the neuron leading to apoptotic death. But was this binding event specific or nonspecific and how can I characterize the antigen?

Most puzzling was how to characterize this apparent autoantibody. An autoantigen is an antigen of the body's own cells that appears similar to other antigens and evokes an antibody response. But was it a specific antigen like a glutamate receptor 3 as in RE, or a nonspecific neuronal antigen that happens to bind as described in the processes of automimicry? The purpose of the new ProteinChip microarray technology for automated analysis of proteins is to try and capture this autoantibody in an attempt to characterize this autoantigen.[10]

I decided to analyze five AD subjects, five normal, nondemented, age-matched subjects, and five lupus subjects on human protein microarray chips that were specially designed to have over 5000 proteins related to autoimmune diseases using the ProteinChip microarray technology. The goal was to find autoantibodies that form the basis for diagnostic, prognostic, and disease progression biomarkers for the two human disease states. These potential biomarkers could be used separately to distinguish disease from normal individuals or in a panel of several biomarkers. After running the assays, too much data were generated and trying to sift through the pages of data seemed fruitless. Several key protein signatures were identified but improvements were needed. A biostatistician analyzed the data and designed special algorithms to manage the data to determine correlations to the subjects.[10] The algorithms typically used to mine DNA microarray data were quite novel in the proteomic field and they helped identify the proteomic patterns in AD that were unique from lupus, and from control serums. Even though it was planned to rerun those assays, it was worth assembling a paper to report these algorithms that were critical for a path forward.[10]

By using the patterns of selected autoantibodies, the controls were distinguishable from the AD and lupus serums, but the goal was to also determine if there was a single AD autoantibody that could be discovered as identified in other CNS autoimmune diseases. As to the autoimmunity aspect of AD, it remains to be determined how significant a role it plays in the neuropathology.

REFERENCES

1. D'Andrea MR. Add Alzheimer's disease to the list of autoimmune diseases. *Med Hypotheses.* 2005;64(3):458–463.

2. D'Andrea MR. Evidence linking autoimmunity to neuronal cell death in Alzheimer's disease. *Brain Res.* 2003;982(1):19–30.

3. Bien CG, Granata T, Antozzi C, et al. Pathogenesis, diagnosis and treatment of Rasmussen encephalitis: a European consensus statement. *Brain.* 2005;128:454–471.

4. Takei H, Wilfong A, Malphrus A, et al. Dual pathology in Rasmussen's encephalitis: a study of seven cases and review of the literature. *Neuropathology.* 2010;30(4):381–391.

5. Minerva. *BMJ.* 2005;330:738.

6. D'Andrea MR. Evidence that the immunoglobulin-positive neurons in Alzheimer's disease are dying by the classical complement pathway. *Am J Alzheimers Dis Other Demen.* 2005;20(3):144–150.

7. Kim YS, Kim SS, Cho JJ, et al. Matrix metalloproteinase-3: a novel signaling proteinase from apoptotic neuronal cells that activates microglia. *J Neurosci.* 2005;25(14):3701–3711.

8. Liphaus BI, Liss MHB. The role of apoptosis proteins and complement component in the etiopathogenesis of systematic lupus erythematosus. *Clinics (Sao Paulo).* 2010;65(3):327–333.

9. D'Andrea MR. Immunoglobulin-positive neurons in Alzheimer's disease are dying via the classical, antibody-dependent, complement pathway. US Patent us 20060024753 A1. 2006.

10. Lubomirski M, D'Andrea MR, Belkowski SM, Cabrera J, Dixon JM, Amaratunga D. A consolidated approach to analyzing data from protein microarrays with an application to immune response profiling in humans. *J Comput Biol.* 2007;14(3):96–105.

CHAPTER 14

The BBB and BRB in AD

… retina as a harbinger of AD …

A dysfunctional BBB is a critical pathological event leading to the anomalous presence of Igs and excessive amounts of Aβ42 into the brain. In 2001, I was trying to imagine how a clinician could assess the integrity of the BBB in subjects, perhaps by some imaging modality using a tracer to see if it penetrates from potential leaky vessels into the brain.

I strongly feel that if a person has a dysfunctional BBB, it is logical to propose that there may also be an association with other cardiovascular disease factors such as hypertension, stroke, diabetes, and high cholesterol. Interestingly, these are all risk factors of AD. But how can you determine or assess the integrity of the BBB? Also it's not clear if dysfunctional BBB occurs in stages to first leak smaller proteins such as Aβ42 into the brain, and then larger proteins such as Igs. If true, then I wonder if the initial pathological events in the AD brain are related to excessively uncontrolled entry of Aβ42 into neurons that die and release factors perhaps through the reactive inflammatory cells that make the BBB even more porous that allows the larger immunoglobulins to wreak their havoc.

Since vascular leakage is a normal physiological response to histamines, viruses, and infections, I wondered if they might also play a role in BBB health. For example, it is possible that a viral or bacterial infection could trigger the BBB to be more permeable to allow antibodies into the brain that would cause a subject with dementia or MCI to deteriorate further. It has been well reported that certain biochemical and physiological changes such as increased levels of inflammatory cytokines, hypoxia-inducible factor-1α, and matrix metalloproteinase-9 in neurological conditions could cause BBB disruption and hence vascular leakage.[1] Is the pathology of the BBB from the inside-out (meaning factors in the brain cause BBB leakage), or from the outside-in (meaning factors in the vasculature cause BBB leakage)?

Interestingly, the need to regulate the entry of vascular components into the brain also exists in other areas of the body, such as in the retina where

Bursting Neurons and Fading Memories. http://dx.doi.org/10.1016/B978-0-12-801979-5.00014-X

there is a blood–retina barrier (BRB). In 2004, I asked my optometrist how much detail you could really see when examining my retina after pupil dilatation. She said she could identify tiny hemorrhages, or tiny breaks in the retina vasculature. I shared my thought that perhaps the retina is the "window" into the brain, or more specifically, "the window of AD" and if the retina had issues, one can propose that the BBB may also have issues.

BRB, A VASCULAR HARBINGER

I began to do extensive literature searches on the association of BBB and BRB, retina pathology and AD, and glaucoma and AD. To my surprise, there were earlier data suggesting that the BRB can be dysfunctional in eye pathologies[2,3] and that there is an association with vascular diseases, which is again a risk factor for AD.[4] In fact, many reports support the notion that AD may actually be more a vascular disease than a neurodegenerative one.[5] De la Torre[5] provided a compelling argument while surveying published literature that revealed no evidence that amyloid deposition is neurotoxic in human beings or that it results in neurodegenerative changes involving synaptic, metabolic, or neuronal loss in human or transgenic mouse brains. By contrast, the data supporting AD as a primary vascular disorder are more convincing, which come from epidemiological, neuroimaging, pathological, and clinical studies.[5] It was concluded in the paper that endothelial dysfunction is an early event in AD patients,[6] further supporting my model.

Endothelial damage may actually be the primary event on BBB and BRB dysfunction, suggesting that the primary pathological event may occur from outside the brain. Endothelial damage is also a primary event in diabetic retinopathy, as BRB breakdown precedes pathological retinopathy in diabetes.[7–9] It has been similarly suggested through animal transgenic AD mouse models that vascular pathologies "precede" the presence of plaques and cognitive impairments.[10] In addition to the detection of amyloid in the cerebrovasculature, which is particularly present in the leptomeningeal and cortical arteries, resulting in cerebral amyloid angiopathy, it was also determined that amyloid is targeted to the vasculature in a mouse model of hereditary cerebral hemorrhage with amyloidosis.[10,11] Also, the Tg-SwD1 mice (transgenic mouse expressing neuronal Aβ precursor protein harboring the Swedish and Dutch/Iowa mutations) display early onset and robust accumulation of Aβ in the brain with a high association with isolated cerebral microvessels that are highly associated with occasional signs of microhemorrage.[12]

So, can you evaluate BRB integrity through a routine eye exam? Apparently so, as there are also data on detailed eye examinations that can reveal BRB dysfunction, which can lead to albumin in the vitreous humor of the eye.[13] For example, all people with diabetes (both type 1 and type 2) are at risk of diabetic retinopathy, which is the reason for a comprehensive eye exam at least once a year for diabetic patients. Between 40% and 45% of Americans diagnosed with diabetes have some stage of diabetic retinopathy that often has no early warning signs, and if left unchecked by an eye care professional, results in small hemorrhages over time (Figure 14.1). Based on the Optomap® website (http://www.optos.com/en-US/Professionals/Image-library/Color-fundus-images/Diabetic-retinopathy/), their scanning system can present images of leaking blood vessels, pale, fatty deposits on the retina, which are themselves signs of leaking blood vessels. Perhaps this kind of technology could be used to assess the integrity of the retina blood vessels in patients with AD and MCI diagnoses to validate this hypothesis.

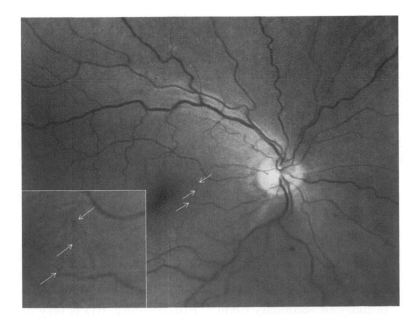

Fig. 14.1. Optomap® 100° retinal image of a patient with diabetic retinopathy. This image demonstrates the optic nerve and macula in a patient's right eye. Small hemorrhages (arrows) can be visualized around the macula (http://www.optos.com/en-US/Professionals/Image-library/Color-fundus-images/Diabetic-retinopathy/).

Beyond the current standard fundus photography, a promising non-invasive method of optical coherence tomography was presented as a way to quantify retinal thickness.[14] Microaneurysm counts, assessment of length and diameter of retinal vessels, and computerized quantification of all pathological elements may also be useful as diagnostic tools and/or efficacy end points.[14] Other nonimaging studies that attempt to assess the permeability of the BRB and BBB have been proposed and include detection of sucrose and albumin. For example, sucrose was used to determine the permeability of the BRB and BBB simultaneously using an intravenous injection in a rat.[15]

There is also a positive correlation between retinal pathology and AD.[16] AD patients often exhibit poor vision and others show visual signs of impairment.[17–19] These clinical manifestations are also supported by regional neuron loss and glial changes in the ganglion cell layer of the retina[20]; however, no explanation was given to the causes of these losses of neurons. I suspect the $\alpha 7$ receptor and Aβ42 may be involved. It was of interest that, despite extensive neuronal loss, no neurofibrillary tangles (NFT) were observed in the retina as shown also for the visual cortex, suggesting that neuronal loss could occur without NFT formation.[20] It was interesting to note that the neuronal loss in the AD eye was different from the neuronal loss from patients with glaucoma.[20] Although there is a decrease in the neurons of the eye during normal aging, the decrease of neurons in the AD eye is well outside the normal range and was not correlated with age, in contrast to the normal.[20]

It is my belief that pathological data obtained by high-resolution analysis of the BRB could indirectly assess the integrity of the BBB, and BBB dysfunction can predict AD, that is, I see the retina as a harbinger of AD through the integrity of the BBB. This is only half of the story as I believe there is at least a two-level paradigm for the causes of AD that starts with the BBB dysfunction leading to the unregulated, and certainly unwelcomed, presence of vascular components such as amyloid and autoantibodies.[21,22] Blood–brain dysfunction is also reported in Binswanger's disease (BD), where serum-derived immunoglobulins have been observed in the brain, although it was unclear if BBB dysfunction was a primary or secondary event.[23] Furthermore, BD is very much a dementia disease and has at least two of the following criteria: hypertension (75%) or known systematic vascular disease, evidence of cerebrovascular diseases (60%), and/or subcortical brain dysfunction.[24]

Autoimmunity is also common in retina pathologies. For example, serum autoantibodies to optic nerve head glycosaminoglycans have been reported in patients with glaucoma.[25] Furthermore, evidence suggests that autoimmune damage to the optic nerve in glaucoma may occur directly by autoantibodies to heat shock protein 27 or indirectly by way of a "mimicked" autoimmune response to a sensitizing antigen to heat shock protein 60 or rhodopsin, which in turn injures retinal ganglion cells.[26] The presence of autoantibodies to neuron-specific enolase was also detected in glaucoma patients that cause retinal dysfunction in vivo.[27] Although it was not mentioned in this article, it is believed that the only way antibodies can pathologically affect the retina is through a BRB dysfunction. However, they similarly claim that antibody avidity, affinity, and specificity, as well as antibody class, will determine the extent of the pathology.

A positive relationship between diabetes, aging, dementia, and stroke has also been reported.[28–32] Diabetes was presented as a probable risk for Alzheimer's disease mainly through the cerebrovascular disease causes,[33] thereby strengthening the association of AD, diabetes, and the BBB. It is also interesting that successful diabetes treatment produces significant cognitive improvement.[34,35]

It is unclear whether small retina changes may be associated with any abnormal clinical features, but work could be "focused" to see if there are subtle changes in the retina vasculature and if they correlate to systematic vascular health and to changes in the BBB function that could lead to cognitive impairment. I feel that it will be invaluable data to collect at perhaps minimal costs, and the rewards could be astronomical. Also, I would explore the frequency of subtle retinal issues in MCI patients as another possible prognostic marker for developing AD.

IN VIVO BBB SUPPORT

With the publications I had up to this point I was disappointed that more interest was not generated. I designed another study that could validate my findings. In a series of experiments, mice were injected with fluorescent Aβ42 through the tail vein, and a few other mice were injected with the pertussis toxin, which is a known toxin used to erode the BBB. The goal was to unequivocally show that Aβ42 can get into the brain through a BBB leak, and will get into the neurons over time. In only an hour, the fluorescently labeled Aβ42 crossed the BBB of pertussis toxin-treated mice and entered the brain tissue as vascular-associated diffuse plaques and also into populations of neurons (Figure 14.2).[36]

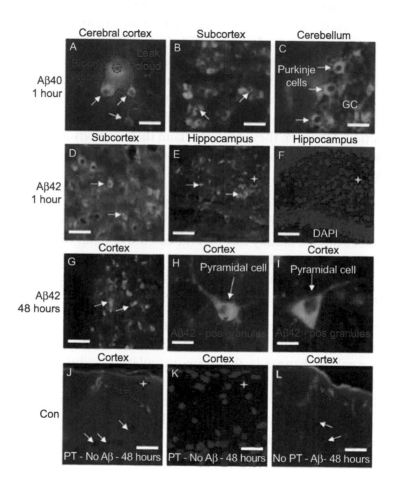

Fig. 14.2. *Blood-borne green fluorescent Aβ42 and Aβ40 cross the blood–brain barrier of PT-treated mice and enter into the brain tissue and bind selectively to neurons (but not glial cells) in the cerebral cortex, subcortex, hippocampus, and cerebellum. In all images, free-standing arrows designate neurons. (A–C) Within 1 hour postinjection, green fluorescent Aβ40 leaks from local vessels (red dotted line) and binds to the surfaces of neurons. Neurons positioned outside of the leak zone show little or no labeling (yellow arrow). (D–F) Neurons with bound green fluorescent Aβ42 are abundant in the indicated brain regions, and comparison of green fluorescent and blue-DAPI (nuclei) image pairs of the same section reveals a preferential binding of Aβ42 on the large neurons occupying the region enclosed by the dentate gyrus. "+" added for orientation. (G) At 48 hours postinjection, labeled neurons are still abundant. (H, I) Some of the larger pyramidal neurons show small bright, green fluorescent Aβ42-positive granules in the basal portion of the perinuclear cytoplasm. (J, K) Basal level of autofluorescence in the cerebral cortex demonstrated in a mouse treated with only pertussis toxin. (L) Mouse given saline in lieu of PT and subsequently treated with green fluorescent Aβ42 demonstrates confinement of fluorescence in the context of an intact BBB. GC, granular cell; PT, pertussis toxin. Scale bar equals 20 μm (I), 30 μm (H), 60 μm (A, C, L, J, K), 100 μm (B, D, L), and 150 μm (E–G). (Courtesy of Brain Res. 2007;1142:223–236.)*

This was proof that Aβ42 can penetrate the brain through a dysfunctional BBB and once in the brain, the Aβ42 will internalize in neurons; therefore, although the neurons are dying from the "inside-out," the true general way to describe the pathological events in the brain is from the outside of the brain. The follow-up study would be a continuation to show that over time these fluorescently labeled Aβ42 neurons die to form dense-core, inflammatory plaque. Finally, for additional absolute certainty, the study should track the behavior of these mice in validated learning models such as the water maze test with well-executed controls.

REFERENCES

1. Wang Z, Meng CJ, Shen XM, et al. Potential contribution of hypoxia-inducible factor-1alpha, aquaporin-4, and matrix metalloproteinase-9 to blood–brain barrier disruption and brain edema after experimental subarachnoid hemorrhage. *J Mol Neurosci.* 2012;48(1):273–280.

2. Krogsaa B, Lund-Andersen H, Parving HH, Bjaeldager P. The blood–retinal barrier permeability in essential hypertension. *Acta Ophthalmol.* 1983;16(4):541–544.

3. Vinores SA. Assessment of blood–retinal barrier integrity. *Histol Histopathol.* 1995;10:141–154.

4. Tatton WT, Chen D, Charmers-Redman R, Wheeler L, Nixon R, Tatton N. Hypothesis of a common basis for neuroprotection in glaucoma and Alzheimer's disease: anti-apoptosis by alpha-2-adrenergic receptor activation. *Surv Ophthalmol.* 2003;48(1):S25–S37.

5. De la Torre JC. Is Alzheimer's disease a neurodegenerative or a vascular disorder? Data, dogma, and dialectics. *Lancet Neurol.* 2004;3(5):270.

6. Borroni B, Volpe R, Martini G, et al. Peripheral blood abnormalities in Alzheimer disease: evidence for early endothelial dysfunction. *Alzheimer Dis Assoc Disord.* 2002;16(3):150–155.

7. Cunha-Vaz JG. Studies on the pathology of diabetic retinopathy. *Diabetes.* 1983;32(2):20–27.

8. Vinores SA, Kuchle M, Mahlow J, Chiu C, Green WR, Campochiaro PA. Blood–ocular barrier breakdown in eyes with ocular melanoma. *Am J Pathol.* 1995;147(5):1289–1297.

9. Vinores SA. Localization of blood–retinal barrier breakdown in human pathologic specimens by immunohistochemical staining for albumin. *Lab Invest.* 1990;62(6):742–750.

10. Su GC, Arendash GW, Kalaria RN, Bjugstad KB, Mullan M. Intravascular infusions of soluble β-amyloid compromise the blood–brain barrier, activate CNS glial cells and induce peripheral hemorrhage. *Brain Res.* 1999;818(1):105–117.

11. Herzig MC, Winkler DT, Burgermeister P, et al. Aβ is targeted to the vasculature in a mouse model of hereditary cerebral hemorrhage with amyloidosis. *Nat Neurosci.* 2004;7(9):954–960.

12. Davis J, Xu F, Deane R, et al. Early-onset and robust cerebral microvascular accumulation of amyloid beta-protein in transgenic mice expressing low levels of a vasculotropic Dutch/Iowa mutant form of amyloid beta-protein precursor. *J Biol Chem.* 2004;279(19):20296–20306.

13. Wong TY. Is retinal photography useful in the measurement of stroke risk?. *Lancet.* 2004;3:179.

14. Lund-Andersen H. Mechanisms for monitoring changes in retinal status following therapeutic intervention in diabetic retinopathy. *Surv Ophthalmol.* 2002;47(S2):S270–S277.

15. Ennis SR, Betz AL. Sucrose permeability of the blood–retinal and blood–brain barriers. *Invest Ophthalmol Vis Sci.* 1986;27:1095–1102.

16. Blanks JC, Torigoe Y, Hinton DR, Blanks RH. Retinal pathology in Alzheimer's disease I. Ganglion cell loss in foveal/parafoveal retina. *Neurobiol Aging.* 1996;17(3):377–384.

17. Cogan DG. Visual disturbances with focal progressive dementing disease. *Am J Ophthalmol.* 1985;100:68–72.

18. Cronin-Golomb A, Sugiura R, Corkin S, Growdon JH. Incomplete achromatopsia in Alzheimer's disease. *Neurobiol Aging.* 1993;14(5):471–477.

19. Katz B, Rimmer S. Ophthalmologic manifestations of Alzheimer's disease. *Surv Ophthalmol.* 1989;34:31–43.

20. Blanks JC, Schmidt SY, Torigoe Y, Porrello KV, Hinton DR, Blanks RH. Retinal pathology in Alzheimer's disease II. Regional neuron loss and glial changes in GLC. *Neurobiol Aging.* 1996;17(3):385–395.

21. D'Andrea MR. Evidence that the immunoglobulin-positive neurons in Alzheimer's disease are dying by the classical complement pathway. *Am J Alzheimers Dis Other Demen.* 2005;20(3):144–150.

22. D'Andrea MR. Evidence linking autoimmunity to neuronal cell death in Alzheimer's disease. *Brain Res.* 2003;982(1):19–30.

23. Akiguchi I, Tomimoto H, Suenaga T, Wakita H, Budka H. Blood–brain barrier dysfunction in Binswanger's disease; an immunohistochemical study. *Acta Neuropathol.* 1998;95(1):78–84.

24. Olsen CG, Clausen ME. Senile dementia of Binswanger's disease. *Am Family Phys.* 1998;58(9):2068–2074.

25. Tezel G, Edward DP, Wax MB. Serum autoantibodies to optic nerve head glycosaminoglycans in patients with glaucoma. *Arch Ophthalmol.* 1999;117(7):917–924.

26. Wax MB, Yang J, Tezel G. Serum autoantibodies in patients with glaucoma. *J Glaucoma.* 2001;10(S1):S22–S24.

27. Maruyama I, Maeda T, Okisaka S, Mizukawa A, Nakazawa M, Ohguro H. Autoantibody against neurons-specific enolase found in glaucoma patients causes retinal dysfunction in vivo. *Jpn J Ophthalmol.* 2002;46(1):1–12.

28. Tariot PN, Ogden MA, Cox C, Williams TF. Diabetes and dementia in long-term care. *J Am Geriatr Soc.* 1999;47(4):423–429.

29. Mortel KF, Wood S, Pavol MA, Meyer JS, Rexer JL. Analysis of familial and individual risk factors among patients with ischemic vascular dementia and Alzheimer's disease. *Angiology.* 1993;44(8):599–605.

30. Skoog I. Risk factors for vascular dementia: a review. *Dementia.* 1994;5:137–144.

31. Kurita A, Katayama K, Mochio S. Neurophysiological evidence for altered higher brain functions in NIDDM. *Diabetes Care.* 1996;19(4):360–364.

32. Ott A, Stolk RP, Hofman A, van Harskamp F, Grobbee DE, Breteler M. Association of diabetes mellitus and dementia: the Rotterdam study. *Diabetologia.* 1996;39(11):1392–1397.

33. Messier C. Diabetes, Alzheimer's disease and apolipoprotein genotype. *Exp Gerontol.* 2003;38:941–946.

34. Gradman TJ, Laws A, Thompson LW, Reaven GM. Verbal learning and/or memory improve with glycemic control in order subjects with non-insulin-dependent diabetes mellitus. *J Am Geriatr Soc.* 1993;41(12):1305–1312.

35. Meneilly GS, Cheung E, Tesier D, Yakura C, Tuokko H. The effect of improved glycemic control on cognitive functions in the elderly patient with diabetes. *J Gerontol.* 1993;48(4):M117–M121.

36. Clifford P, Zarrabi S, Siu G, et al. Abeta peptides can enter the brain through a defective blood–brain barrier and bind selectively to neurons. *Brain Res.* 2007;1142:223–236.

CHAPTER 15

"Inside-Out" in the Field

… epithelial cells can be a large source of amyloid …

How does the new hypothesis fit into your understanding of what causes AD? If you consider the previous hypotheses presented, only a few overlap, but, and as far as I can tell, none directly invalidate my findings and explanations. All of these hypotheses certainly include explanations of neuronal death, but fail to logically explain how death occurs with as much evidence as has been presented in this book. I sincerely hope that this hypothesis helps to align, unify, and even help explain many of the various hypotheses about the causes of AD. In this chapter, I will try to place the "inside-out" hypothesis in the context of some of these other proposals.

NEURONAL DEATH BY AMYLOID (VIA VASCULAR ISSUES)

I have no doubt that amyloid plays an integral pathological role in neuronal death, however not as described in the amyloid hypothesis. Since 1998, I've never seen evidence to support the notion that amyloid, in particular Aβ42, aggregates outside the neurons to kill neighboring neurons, nor have I seen evidence that Aβ42 is directly toxic, only guilty because of the dysfunctional BBB allowing unregulated copious amounts of Aβ42 into the brain to eventually contribute to neuronal death by overindulging.

The real problem with Aβ42 is how readily it is able to enter the brain when the typical regulation, in this case the BBB, fails. The fact that the α7 receptor has such a high binding affinity to Aβ42 suggests that high levels of Aβ42 are not typically present in the brain. As previously noted, Aβ42 is present in normal neurons, and so it is the uncontrollable amounts of vascular-derived Aβ42 that enter the AD brain that is the core concern. It is the dysfunctional BBB that is the problem that leaks Aβ42 into the brain as further evidenced by the presence of diffuse amyloid plaques. Over time, the overloaded neurons burst leaving indigestible neuronal debris in their stead.

Bursting Neurons and Fading Memories. http://dx.doi.org/10.1016/B978-0-12-801979-5.00015-1

I am also supportive of hypotheses that directly or indirectly impli-
cate vascular risk factors, including advancing age, cholesterol, and, of
course, the vascular hypothesis, as catalysts for dysfunction in the BBB
where AD begins.

However, what is the source of this amyloid, and how does amyloid
make its way out of the vasculature? As to the former, epithelial cells of the
thyroid and small intestine can be a large source of the types of amyloid
associated with AD.[1,2] Interestingly a high-fat diet, another AD risk factor, simulates amyloid production in the small intestines thereby increasing
vascular levels of amyloid.[2] As to the latter, several cardiovascular risk factors also increase the risk of AD (Chapters 12 and 14). Amyloid, specifically $A\beta42$, is about 4.5 kDa and is significantly smaller ($\sim$33 times) than the
typical antibodies that are about 150 kDa, and may indicate that amyloid
could enter earlier and more often and then as the neurons engorge themselves and eventually die, further eliciting the inflammatory responses that
could lead to additional BBB dysfunction that allows even larger vascular
components into the brain; what a serious spiral out of control!

NEURONAL DEATH BY ApoE4

The APOE gene, the somewhat leading genetic risk factor for AD, has been
largely obscured and demoted. Like many other nonamyloid hypotheses,
ApoE had been put on the back burner due to the craze of studying and
funding amyloid studies.[3] Now that amyloid-based studies have failed,
researchers are developing drugs aimed at the ApoE4 protein independent
of amyloid. When neurons are under stress, they make ApoE as part of a
repair mechanism. The ApoE4 form is broken down into toxic fragments
that damage the cell's energy factories, the mitochondria, while altering
the cell skeleton. A planned trial will investigate whether a low dose of
pioglitazone, already approved for certain patients with type 2 diabetes,
can delay AD onset, since preclinical and small-scale human studies
suggest it may prevent or reverse AD-related pathology and symptoms.[4]

NEURONAL DEATH BY INFLAMMATION

I certainly make the case that inflammation plays a significant role in
AD symptoms and pathological deterioration, which is why I believe
its suppression plays an equally important role. Minimizing or even

eliminating inflammation may help mitigate further neuronal damage and death, although it will not prevent the initial cell death. My model is in complete contrast to the chronic inflammation model, whereby cellular stress to aging neurons leads to hyperphosphorylated tau, and impaired axonal transport leading to the accumulation of APP in the axonal compartments as large "swelling." These swellings form the diffuse plaques, as the microglia struggle to clear the dystrophic neurites and debris.[5] This diffuse plaque eventually leads to the formation of the senile plaque, as the nearby neuron becomes embodied with neurofibrillary tables due to caspase activation. This model does not account for intracellular Aβ42 nor does it account for most of the evidence provided in this book.

NEURONAL DEATH BY TAU

As I noted at the Neuroscience meeting in Miami, Aβ42 induced rapid phosphorylation of tau via the α7 receptor, which was subsequently published by my collaborators.[6] And so the connection is obvious. I cannot speak to tau only that I believe that tau, specifically, paired helical filaments and neurofibrillary tangles, is the consequence of overloaded intracellular amounts of Aβ42 and may also contribute to cell death. Years ago, I performed double IHCs using antibodies to the various phosphorylated forms of tau identified in paired helical filaments, with Aβ42 and with the α7 receptor, and it was rare to see α7 receptor-positive neurons that were also positive for the tau forms, and were Aβ42-negative. However, all of the α7 receptor-positive neurons were Aβ42-positive, and I did observe a smaller subpopulation with the tau forms, including the tangles, but most were Aβ42-positive with normal tau suggesting the any tau pathology is a downstream consequence, although this was admittedly a personal observation.

NEURONAL DEATH BY AUTOANTIBODIES

This hypothesis remains to be challenged. I hope in the upcoming years, the presence of neuronal-specific autoantibodies will indicate the presence of neuronal death as a biomarker to which the body already initiated a classical inflammatory reaction. It will be equally important to keep these large proteins from entering the brain, via a dysfunctional

BBB. How else can you explain the presence of Igs in the brain that appear to be associated with pending cell death? I would be most interested in understanding why some neurons were Ig-positive and some nearby were Ig-negative, which I suppose is based on the uncharacterized autoantigen.

MISCELLANEOUS NOTES

Does everyone with a dysfunctional BBB have an increased risk of AD? I would say so, but like many other diseases including cancer, it is the combination of several risk factors that leads to diseases. For example, consider the autoimmune component of AD: if you have a dysfunctional BBB and you do not have antineuronal autoantibodies, then you will not exhibit signs of AD. Similarly, if you have antineuronal antibodies, and your BBB is intact and normal, you will not likely exhibit signs of AD, but having those antibodies and a dysfunctional BBB would be a recipe for pending central nervous system issues that could range from dementia to AD depending on the type and characteristics of antineuronal autoantibody. The only way to verify would be to test this possibility.

The "inside-out" hypothesis is not the ultimate answer to understanding AD neuropathology but is a logical one. I have been able to systematically explain my observations and compare them with the currently embraced amyloid hypothesis. It is clear that neuronal cells are dying, which explains loss of brain function as well as the decrease in brain size leading to AD. The neurons are dying by necrosis and apoptosis, perhaps inflicted by Aβ42 and Ig, respectively, but which comes first? If the pathology begins from outside the brain, then early stages of BBB dysfunction could begin by the passive leakage of Aβ42 into the brain. Neurons are not accustomed to unregulated amounts of Aβ42 and overingest leading to self-inflicted cell death. To continue, as the cell dies and the process of inflammation is triggered, perhaps these cells release factors to cause the BBB to become even leakier and the vascular pathology becomes more severe to allow larger-sized vascular elements, such as the Igs, into the brain. To make matters even worse, nearby neurons are fatally affected by the cataclysmic, exploding neurons as their processes are caught in the wake of the lytic event. A montage of seemingly endless pathological activities including a defective security barrier, exploding cells, engaging inflammatory cells, loss of neighboring cells, and autoimmune effects leads to brain-wide failure.

In summary, the "inside-out" hypothesis can easily address the following that the amyloid hypothesis cannot:

1. Why are there no dense-core plaques in the molecular layer of the brain? No neurons are present.
2. Why is intracellular Aβ42 critical? Too much leads to cell death, therefore need to inhibit internalization.
3. Why are there nuclei in the center of some dense-core plaques? It is the neuronal remnant, not all unless you typically perform serial sectioning to catch that egg yolk.
4. Why is lipofuscin in dense-core plaques? Remains where the neuron died as proteolytic neuronal debris.
5. How is there an inverse relationship between the numbers of dense-core plaques and neurons? Since they are one in the same, meaning dense-core plaques are dead neurons, as the number of dense-core plaques increases, the number of neurons decreases.
6. How can the lack of MAP-2 immunolabeling in dense-core plaques be explained? As the neurons burst, the released enzymes digest proteins sensitive to digestion.
7. How to explain the presence of neuronal cellular components in dense-core amyloid plaques? Those neuronal components that are resistant to enzymatic digestion are present as neuronal debris when the cell died.
8. How to explain the fairly narrow range of dense-core plaque sizes? As you descend the neuronal layers in the brain, the neuronal sizes increase as do the relative sizes of the dense-core plaques; again you have to catch the spherical plaque in the middle section. The actual sizes may be dependent on the finite activity of the released proteolytic enzymes that digest proteins in a radial manner in the brain.
9. How to explain the origin of the diffuse plaques? Derived from amyloid leaking in through the BBB, and can also form from degenerating Purkinje dendritic fibers, essentially inconsequently to the neuropathology of AD, and removing this type of extracellular amyloid will not improve behavior or improve memory.

10. How to explain that some plaques are associated with inflammatory cells, while others are not? Depends on the plaque type: plaques formed from dying cells (neurons, astrocytes) are associated with inflammatory cells (gliosis), while plaques formed from leakage (diffuse types) are not associated with gliosis.

11. How to explain the spherical shape of dense-core plaques? Radial dispersion of lysosomes, perhaps based on the time of activity and the inability to continue to penetrate the brain tissue further than the typical range; remember to perform serial sections to ensure you measure the spherically shaped plaque at the equator.

12. How to explain the random morphologies of the diffuse plaques? Random vascular leaking events with widely observed sizes and shapes.

13. How to explain the presence of antineuronal antibodies in the AD blood? AD also presents an autoimmune pathology and I believe that the dead neuronal debris percolated through the brain cerebral spinal fluid to eventually provoke an inflammatory response to produce antibodies, or if it is automimicry, then it is unfortunate that the similar antibodies accidentally found their way into the CNS to happen to bind to similar antigens.

14. How to explain the presence of $A\beta42$ in normal neurons? $A\beta42$ is not toxic and is a normal physiological protein.

15. How to explain the failure of the current clinical programs to meet their efficacy end points to thereby improve cognitive function? First, I presume that the preclinical data used to advance the project into clinical trials were based on the removal or reduction of amyloid plaque load. And based on the amyloid cascade hypothesis, removal of extracellular amyloid is the key to curing AD. However, the most important concept of my hypothesis is that removing either plaque type is a fruitless efficacy end point. If you remove the dense-core, senile amyloid plaque, it doesn't matter as the neuron is already dead, and if you remove the diffuse plaque types, it will not matter either as they are not the pathological type of plaque. Removing plaques is a superfluous exercise based on the amyloid cascade hypothesis and will not treat the disease.

16. How to explain the known cardiovascular risk factors for having AD? AD is a cardiovascular disease and dysfunction of the BBB

that leads to the unregulated entry of blood components such as amyloid and immunoglobulins.

17. Previous clinical studies to suppress inflammation have also failed; how can you explain? Although inflammation may play a role in "secondary" neuronal death, amyloid overloading and autoimmune-dependent cell death are primary events. Alleviating the inflammation cascade can only keep the pathology from making matters worse, but it will never reverse the course of the death of those neurons that remain as neuronal debris covered as glial scars.

I hope you learn from my experiences to trust your observations in the face of overwhelming contradictive hypotheses. Perhaps this story will make you a bit more alert, open-minded, and will help make the difference to reset the direction of AD research, as it's critical to understand that in time, what begins at the bench ends up at the clinic.

In sum, two of the primary messages stem from unexpected detection of intracellular Aβ42 and intracellular Ig in neurons that are based on a dysfunctional BBB. It is my hope that these two discoveries as well as the known BBB issues, along with your help, will guide AD research into new successful areas of diagnosis, prognosis, treatment, and, ultimately, prevention.

REFERENCES

1. Schmitt TL, Steiner E, Klingler P, Lassmann H, Grubeck-Loebenstein B. Thyroid epithelial cells produce large amounts of the Alzheimer beta-amyloid precursor protein (APP) and generate potentially amyloidogenic APP fragments. *J Clin Endocrinol Metab*. 1995;80(12):3513–3519.

2. Galloway S, Jian L, Johnson R, Chew S, Mamo JC. Beta-amyloid or its precursor protein is found in epithelial cells of the small intestine and is stimulated by high-fat feeding. *J Nutr Biochem*. 2007;18(4):279–284.

3. Spinney L. The forgetting gene. *Nature*. 2014;501:26–28.

4. Sato T, Hanyu H, Hirao K, Kanetaka H, Sakurai H, Iwamoto T. Efficacy of PPAR-γ agonist pioglitazone in mild Alzheimer disease. *Neurobiol Aging*. 2011;32(9):1626–1633.

5. Krstic D, Knuesel I. Deciphering the mechanism underlying late-onset Alzheimer disease. *Nat Rev Neurol*. 2013;9:25–34.

6. Wang HY, Li W, Benedetti NJ, Lee DH. Alpha 7 nicotinic acetylcholine receptors mediate beta-amyloid peptide-induced tau protein phosphorylation. *J Biol Chem*. 2003;278(34):31547–31553.

CHAPTER 16

Alzheimer's Disease Tomorrow

… all AD information should be updated …

So how would you prevent AD? You can begin by avoiding the recognized risk factors including obesity, hypertension, diabetes, lack of exercise, and poor diet.[1]

Did the stories in this book alter your opinion? The title of a recent 2014 article stated that AD is "still a perplexing problem."[2] However, I hope this "inside-out" hypothesis provides a logical and unifying explanation to the death of neurons, and the origin of plaques, while helping to align the many other AD hypotheses (e.g., cholinergic, vascular/cholesterol, genetic, inflammatory, and amyloid) and risk factors into one cohesive process of the neuropathology of AD that begins from outside the brain ("outside-in"). This is not a disease initiated in the brain, as currently believed; AD begins at the vascular level.

I am very appreciative that you spent your time reading this collection of observations and hypotheses and if you are in the field looking for a new research project on the bench or at the clinic, I believe I have listed several possible opportunities. I sincerely hope you consider testing some or all of these hypotheses, for this work would have little value without being continually tested and retested to advance AD research, as the cure can only be obtained if the cause is defined.

STATE OF THE AD NATION

Unfortunately, the current state of the AD research field is bleak and without direction. According to the latest study, 99% of Alzheimer's drug trials fail.[3] The deliverables for AD drug development are embarrassing for a field so well-funded. A recent report performed a comprehensive review of all clinical trials underway and showed that there are relatively few drugs in development, the failure rate of AD drug development is 99.6% (years 2002–2012), and the number of drugs continues to decline since 2009.[3] The report further states that according to researchers, in

Bursting Neurons and Fading Memories. http://dx.doi.org/10.1016/B978-0-12-801979-5.00016-3

order to accelerate the drug development process and reduce the need to constantly invent new drugs, there needs to be more repositioning studies. For example, researchers are investigating if bexarotene (Targretin™), an FDA-approved drug to treat skin cancer, can remove a protein buildup in the brains of Alzheimer's patients, as it did in a recent preclinical, animal study. Assuming you have read this book, I think you know my opinion that removing "protein buildup" (assuming they mean amyloid) is an inappropriate efficacy end point; however, there is always hope that the drug may work through yet undefined mechanisms.

So where is the hope? Hope will come from new scientists proposing fresh hypotheses while revisiting previously those dismissed or suppressed.

TARGETING THE α7 RECEPTOR

Back in 2006, I was invited to submit a paper to *Current Pharmaceutical Design*, titled *Targeting the Alpha 7 Nicotinic Acetylcholine Receptor to Reduce Amyloid Accumulation in Alzheimer's Disease Pyramidal Neurons*.[4] As of writing this book, it has just about 100 citations, and yet, I strongly believe this is one of the remaining path forwards to treat and/or prevent AD. We have shown that α7 inhibitors prevent the internalization of Aβ42 in neurons in culture and based on the presence of Aβ42 in smooth muscle cells, I would suspect that α7 inhibitors will also inhibit the endocytosis of Aβ42 in smooth muscle cells thereby preventing vascular plaques and also may minimize leakage of unregulated vascular components into the brain. These data suggest opportunities to reduce intracellular Aβ42 to prevent neuronal death (the inflammatory dense-core plaque) and vascular plaques.

It is especially intriguing when α7 receptor ligands such as nicotine and epibatidine, an analgesic, were able to inhibit Aβ42 binding to the α7 receptor, which may account for the abilities of these compounds to protect against Aβ42-induced cytotoxicity in vitro. Interestingly, in a 2012 clinical trial, nicotine was delivered through a transdermal patch to subjects with MCI for over 6 months; these subjects showed improvement in primary and secondary cognitive measures of attention, memory, and psychomotor speed, and improvements were seen in patient/informant ratings of cognitive function.[5] Although the mechanism leading to the benefit was unclear, based on the earlier chapters in this book, you can hypothesize that nicotine is reducing the amount of Aβ42 into the neurons by competing with Aβ42 for the same α7 receptor, like that of α-bungarotoxin in our previous in vitro work, further suggesting the critical role of the α7 receptor in AD.[6]

TARGETING Aβ42: NOT PLAQUES

Back in 2002, I was invited to submit a publication to *Drug Discovery Research*, titled *Targeting Intracellular Aβ42 for Alzheimer's Disease Drug Discovery*.[7] As noted in this publication, the overproduction of β-amyloid, especially Aβ42, the self-assembling peptide of 42 amino acid residues derived from the proteolytic cleavage of the APP that is found in patients with early onset familial AD and in elderly Down's syndrome patients, strongly suggests that Aβ plays a significant role in the pathogenesis of AD. In this publication I continued to challenge "one of the most embraced models of AD, the amyloid hypothesis." I referred to the sequence of neuronal death found in the "inside-out" hypothesis to emphasize the distinct intracellular Aβ accumulation that contrasts the widely accepted extracellular Aβ deposition theory of amyloid plaque formation in AD pathology. It was cited a sorrowful 19 times and is lost in time, which is the ultimate purpose of this book, to help revitalize the research. We need to target the Aβ42 before it gets into the neurons. I suppose if this work was more embraced, others may have tried to validate or invalidate this hypothesis years ago, but like other lost hypotheses, we need to reconsider that viability of those proposals in an AD field parched for alternative approaches. With that being said, perhaps an effective application of the anti-amyloid antibody bapineuzumab could be to remove amyloid in the bloodstream before it gains entry into the central nervous system via BBB dysfunction. As mentioned, removing plaques is equivalent to cleaning up the mess after the neuron died. If there was any hope to infuse life into this apparently hopeless approach, it would be to stratify or target pre-AD subjects, perhaps those with MCI at high AD risk factors, with specific attention to those with cardiovascular issues such as BBB leakage. If the antibody was able to effectively reduce serum levels of amyloid, specifically Aβ42, then, at most, this could prevent the disease from progressing to AD. Since Aβ42 is a physiologically harmless and perhaps beneficial protein, one must consider inconsequential deleterious effects that may actually be outweighed by the beneficial effects of removing excessive amounts of vascular Aβ42. However, in such cases of MCI and early stages of AD, I would not think that would be a concern.

TARGETING THE BBB

All readers need to appreciate the critical role that the BBB plays in AD, or, generically speaking, the maintenance of a healthy cardiovascular system, a known AD risk factor. So how do you target the BBB to prevent the uncontrolled, unregulated entry of vascular components into the brain? To begin, it's equally important to be aware of potential causes of unregulated BBB leakage. The clinical and prognostic significance of cerebral microbleeds (MBs) manifesting as small chronic brain hemorrhages that were detected using specific magnetic resonance imaging sequences was reviewed in AD.[8] MBs were associated with vascular risk factors such as age and hypertension as well in severe vascular conditions such as ischemic and hemorrhagic stroke. They were also associated with cerebral amyloid angiopathy caused by the accumulation of Aβ on the vessel walls like that presented in Figure 7.8. The presence of MBs in subjects, especially multiple MBs, was associated with slower processing speed and worse executive function. Multiple MBs and retinopathy significantly increased the odds of vascular dementia. AD subjects with multiple MBs had a more severe cognitive impairment and degree of atrophy,[9] but, more importantly, the presence of at least one MB yielded a more than twofold increase in predicting the progression from MCI to dementia, but not a significant risk of non-AD dementia.[10]

It was reported that leakage of the BBB is associated with other neurological disorders, including temporal lobe epilepsy.[11] Following ischemic stroke, the integrity of the BBB can be impaired in cerebral areas distant from the initial ischemic insult, a condition known as diaschisis, leading to chronic poststroke deficits.[12] Interestingly, in the ischemic rat brain model, the late administration of vascular endothelial growth factor (VEGF) enhances angiogenesis in the ischemic brain, improving neurological recovery, and the early administration of VEGF exacerbates BBB leakage. Hence, the controlled regulation of VEGF could be a potentially effective therapeutic strategy aimed at administration of exogenous VEGF to promote therapeutic angiogenesis during the repair process after a stroke and inhibition of VEGF at the acute stage of stroke to reduce the BBB permeability and the risk of hemorrhagic transformation after cerebral ischemia.[13]

Perhaps therapeutics directed to treat or prevent vascular disorders associated with diabetes and hypercholesterolemia could be effective as the treatment of preclinical models of these pathological conditions with darapladib, a selective inhibitor of lipoprotein-associated phospholipase-A2, blocked the progression of atherosclerosis while reducing BBB leakage.[14] Statins have been shown to ameliorate BBB dysfunction resulting from a number of conditions, including diabetes, transient focal cerebral ischemia, and HIV-1.[15–17] The treatment of simvastatin, a natural statin derived from fermentation, was effective in reducing the BBB permeability as measured by Evan's blue dye across the BBB in rabbits fed a cholesterol-enriched diet.[18]

It remains to be determined what other environmental factors also contribute to BBB damage, although some have been investigated.[19] Sustaining a healthy cardiovascular system seems imperative for one's overall well-being, which may begin by having low blood pressure and normal blood chemistry. Future research directed to investigate the association of the integrity of the BBB via imaging examinations with AD may provide yet another important AD risk factor that could then be investigated at the MCI stages of dementia.

TARGETING INFLAMMATION

Clinical trials with anti-inflammatories have produced unsatisfactory results. As described in the "inflammatory cascade," these processes are secondary to the dying neurons, and so, these trials may have had unrealistic expectations. However, such interventions may help to slow down the progression due to the potentially destructive consequences of the inflammatory cells leading to subsequent pathologies. In a recent study that followed 247 AD patients over 13 years, neuroinflammation was independently linked to early death, thereby rapidly progressing the disease.[20] This study suggested that inflammation, not amyloid or tau pathology, is an independent underlying mechanism in AD neuropathology, which supports the inflammatory cascade model.

BIOMARKER DISCOVERY

Over the past 15 years, biomarker discovery has become an integral part of drug discovery as it did for my target validation team at that time. As you may know, the types of biomarkers vary based on the need. For example, some biomarkers are sought to validate target engagement with the study drug, while others are used to stratify target populations. There is no doubt that biomarkers are needed in the AD field but what types would be essential? Again that depends on the need.

Based on the discussion of the cardiovascular system presented in this book, retinal imaging would be an invaluable biomarker to indirectly assess the integrity of the BBB. Earlier upstream pathologies point to BBB dysfunction implying the importance of cardiovascular health that also explains the various cardiovascular risk factors for AD.

Perhaps the levels of vascular amyloid, or specifically Aβ42, might be a risk factor, which alone may appear to not correlate with AD unless coupled with cardiovascular risk factors.

How to assess neuronal death is somewhat determined through clinical testing, but how can you assess neuronal loading? Can you discover a biomarker to assess the health of the neurons affected by the loading of Aβ42 into the neurons? To the latter, probably not as the resolution of imaging is not at the cellular level, but to the former, perhaps a biomarker to assess the early stages of neuronal death before clinical presentation could be discovered. For example, the expression of MAP-2 is missing in areas of the dense-core, senile amyloid plaques due to neuronal lysis. Either the antigen of the MAP-2 is unrecognizable by the primary antibody leading to the lack of IHC labeling or the MAP-2 was digested and is missing in the area of the neuronal debris. If the former, then it is equally possible that fragments of MAP-2 could be detected in the cerebral spinal fluid and/or vascular system as an indication of neuronal death. Possible autoantibodies to fragments of MAP-2 and to other neuronal debris might also provide a means to assess neuronal death as a diagnostic and potentially prognostic biomarker. Perhaps there are markers of BBB dysfunction based on endothelial damage that may be detectable before clinical symptoms are manifested that are not just for AD but also for other CNS-related diseases like that of Rasmussen's disease. In addition, hopefully the autoantigens to which those auto-antibodies bind might help launch an entire new field of AD research.

The autoimmune aspects of AD remain to be characterized and as suggested in an early publication, the goal was to find autoantibodies using the ProteinChip technology that would form the basis for diagnostic, prognostic, and disease progression biomarkers for the two human disease states, AD and lupus.[21] These potential biomarkers may be used separately to distinguish diseased from normal individuals or in a panel of several biomarkers to strengthen the decision-making process. These approaches have since been published in the AD field,[22] while others employ other technologies such as the "antigen surrogate" approach to screen the human blood for disease markers.[23]

This autoantibody profiling might not only help in the diagnosis of AD before the clinical presentation of AD but also help to stratify those subjects with MCI who may be prone to developing AD. Perhaps once the AD autoantibodies are discovered, it may be possible to treat AD in the same way as treating lupus patients, through plasmapheresis. An interesting point is that plasmapheresis is also used to treat patients with myasthenia gravis. In addition, total plasma exchange transiently reduced anti-GluR3 antibody titers and seizure frequency, and improved neurological function in some patients[24,25]; however, since the majority of RE patients show only a transient response, plasmapheresis is regarded as adjunctive therapy in cases of acute deterioration and very frequent seizures. It would still be of great interest to understand the nature, or antigen, of these antineuronal autoantibodies.

ASSESSING BBB INTEGRITY VIA BRB

Assessing BBB leakage directly in preclinical models is typically based on the extravasation of Evan's blue dye, which binds to serum albumin. However, this method is not sensitive enough to detect minor leaks leading to the proposed use of optical imaging to map BBB leakage in a rat thromboembolic stroke model.[26] In a recent newspaper article, there are studies investigating the presence of amyloid in the retina of the eye, and of the first 40 patients in a 200-participant study, retina changes strongly correlated with amyloid plaque development in the brain.[27] It is interesting to see that others are considering the eye as a "window in the AD," but for me, as noted previously, I consider the integrity of the retina as a harbinger of BBB integrity, and that study seems to support indirectly that amyloid in the eye is an indicator of amyloid in the brain. However, it is the presence of a dysfunctional BBB that is observed in the BRB.

There are certainly opportunities to use retinal imaging to indirectly assess the patients' BBB as a risk factor for AD. It remains to be seen if there is association between the health of the retinal vascular system and the BBB, and if the integrity of the BBB correlated with AD. We will know this information only when clinicians begin to include longitudinal retinal scans in their clinical studies since the resolution of such imaging continues to improve to further provide more and more data.

NEW WORK

Recently the discovery of an abnormally high concentration of GABA, an inhibitory neurotransmitter, was found in AD brains. The research showed that the excessively high concentration of the GABA neurotransmitter in these reactive astrocytes is a novel biomarker that the authors hope can be targeted in further research as a tool for the diagnosis and treatment of Alzheimer's disease.[28] After they inhibited the astrocytic GABA transporter to reduce GABA inhibition in the brains of the AD mice, they found that these mice showed better memory capability than those of the control AD mice.

CLOSING STATEMENT

In William Huggins' words: "The riddle of the nebulae was solved."

In my words: "The riddle of the dense-core, senile amyloid plaque is solved." However, since the key to preventing AD begins within the neurons, we have to keep the unregulated entry of vascular-derived amyloid out of the brain.

The "inside-out" hypothesis should change our fundamental understanding about the causes of Alzheimer's disease: that neuronal death comes from within the cell, with neurons releasing their contents out to form the plaque, not the other way around. However, the brain is affected from the "outside-in" by vascular-derived Aβ42, and Igs, or antineuronal antibodies entering the brain without regulation. All AD-related information including diagrams should be updated to incorporate the leaky, dysfunctional BBB, and how this allows the unregulated presence of amyloid and immunoglobulins in the brain that eventually kill neurons. Current images of amyloid plaques aside of neurons are inaccurate and misleading. Cartoons of AD pathology need to first show internalization of amyloid in neurons that over time leads to neuronal lysis to show bursting neurons with surrounding inflammatory cells as the amyloid plaque forms. The subsequent pathological effects of a single bursting neuron could be unimaginable, evident by the lack of MAP-2 immunolabeling and the loss of neuronal cells in that H&E figure, thereby producing a cataclysmic, radial effect damaging the countless number of processes of local and distant neurons located in this "ground zero" area. Updating the images describing the processes of neuronal death and its repercussions on nearby neurons will provide additional push to move the AD research into more successful directions.

As you can see there are many other areas to investigate to help diagnose, treat, and prevent AD. At this time, I can only suggest maintaining a healthy cardiovascular system, and for the brain; a busy and thinking brain translates into a brain that is continually working. Find ways to engage yourself in new learning to stimulate the brain to keep your neurons busy wiring new synapses as a busy neuron will require a constant supply of glucose and oxygen provided by the vascular system. In a recent 2013 article, neuronal brain activity was linked to glucose metabolism.[29] They reported that the brain increases its glucose supply on neuronal activation indicating that the brain supplies itself with energy according to its needs. Regular exercise has been demonstrated to enhance executive cognitive function of selective attention and conflict resolution in a randomized controlled clinical trial in senior women.[30]

Not until the hypotheses in this book are validated can one envision therapies to block Aβ42 from binding to the α7 receptor, remove undesirable anti-neuronal antibodies, or remove excessive Aβ42 in the vasculature. Based on the interpretation of these findings, by the time a patient has AD, it is already too late to recover clinical function that results from dead neurons. Perhaps one can suppress the progression of early to late-stage AD, but, unfortunately, it is impossible to reverse. Prevention must begin at the early stages for subjects with MCI, or earlier, coupled with AD risk/biomarker factors that may include vascular levels of amyloid, assessing the integrity of the BBB directly through high-resolution imaging, or indirectly through retinal scans. Unlike many other cell types in your body, once CNS neurons die, they are not replaceable as "bursting neurons lead to fading memories" and loss of function.

REFERENCES

1. Middleton LE, Yaffe K. Promising strategies for the prevention of dementia. *Arch Neurol.* 2009;66:1210–1215.

2. Chinthapalli K. Alzheimer's disease: still a perplexing problem. *BMJ.* 2014;349:g4433.

3. Cummings JL, Morstorf T, Zhong K. Alzheimer's disease drug-development pipeline: few candidates, frequent failures. *Alzheimers Res Ther.* 2014;6:37–42.

4. D'Andrea MR, Nagele RG. Targeting the alpha 7 nicotinic acetylcholine receptor to reduce amyloid accumulation in Alzheimer's disease pyramidal neurons. *Curr Pharm Des.* 2006;12(6):677–684.

5. Newhouse P, Kellar K, Aisen P, et al. Nicotine treatment of mild cognitive impairment: a 6-month double-blind pilot clinical trial. *Neurology.* 2012;78:91–101.

6. Wang HY, Lee DHS, D'Andrea MR, Peterson PA, Shank R, Reitz A. β-Amyloid1-42 binds to α7 nicotinic acetylcholine receptor with high affinity: implications for Alzheimer's disease pathology. *J Biol Chem.* 2000;275(8):5626–5632.

7. D'Andrea MR, Nagele R, Lee DHS, Wang H-Y. Targeting intracellular Aβ for Alzheimer's disease drug discovery. *Drug Dev Res.* 2002;56:194–200.

8. Martinez-Ramirez S, Greenberg SM, Viswanathan A. Cerebral microbleeds: overview and implications in cognitive impairment. *Alzheimers Res Ther.* 2014;6:33–40.

9. Goos JD, Kester MI, Barkhof F, et al. Patients with Alzheimer disease with multiple microbleeds: relation with cerebrospinal fluid biomarkers and cognition. *Stroke.* 2009;40:3455–3460.

10. Staekenborg SS, Koedam EL, Henneman WJ, et al. Progression of mild cognitive impairment to dementia: contribution of cerebrovascular disease compared with medial temporal lobe atrophy. *Stroke.* 2009;40:1269–1274.

11. van Vliet EA, da Costa Araujo S, Redeker S, van Schaik R, Aronica E, Gorter JA. Blood–brain barrier leakage may lead to progression of temporal lobe epilepsy. *Brain.* 2007;1302:521–534.

12. Garbuzova-Davis S, Haller E, Williams SN, et al. Compromised blood–brain barrier competence in remote brain areas in ischemic stroke rats at chronic stage. *J Comp Neurol.* 2014;522(13):3120–3137.

13. Zhang ZG, Zhang L, Jiang Q, et al. VEGF enhances angiogenesis and promotes blood–brain barrier leakage in the ischemic brain. *J Clin Invest.* 2000;106(7):829–838.

14. Acharyaa NK, Levina EC, Clifford PM, et al. Diabetes and hypercholesterolemia increase blood–brain barrier permeability and brain amyloid deposition: beneficial effects of the LpPLA2 inhibitor darapladib. *J Alzheimers Dis.* 2013;35:179–198.

15. Mooradian AD, Haas MJ, Batejko O, Hovsepyan M, Feman SS. Statins ameliorate endothelial barrier permeability changes in the cerebral tissue of streptozotocin-induced diabetic rats. *Diabetes.* 2005;54(10):2977–2982.

16. Nagaraja TN, Knight RA, Croxen RL, Konda KP, Fenstermacher JD. Acute neurovascular unit protection by simvastatin in transient cerebral ischemia. *Neurol Res.* 2006;28(8):826–830.

17. Lee S, Jadhav V, Lekic T, et al. Simvastatin treatment in surgically induced brain injury in rats. *Acta Neurochir Suppl.* 2008;102:401–404.

18. Jiang X, Guo M, Su J, et al. Simvastatin blocks blood–brain barrier disruptions induced by elevated cholesterol both in vivo and in vitro. *Int J Alzheimers Dis.* 2012;2012:109325.doi: 10.1155/2012/109325.

19. Staekenborg SS, Koedam EL, Henneman WJ, et al. Progression of mild cognitive impairment to dementia: contribution of cerebrovascular disease compared with medial temporal lobe atrophy. *Stroke.* 2009;40:1269–1274.

20. Nagga K, Wattmo C, Zhang Y, Wahland L-O, Palmqvist S. Cerebral inflammation is an underlying mechanism of early death in Alzheimer's disease: a 13-year cause-specific multivariate mortality study. *Alzheimers Res Ther.* 2014;6:41.

21. D'Andrea MR. Immunoglobulin-positive neurons in Alzheimer's disease are dying via the classical, antibody-dependent, complement pathway. US Patent us 20060024753 A1. 2006.

22. Nagele E, Han M, DeMarshall C, Nagele RG. Diagnosis of Alzheimer's disease based on disease-specific autoantibody profiles in human sera. *PLoS One.* 2011;6(8):e23112.

23. Raveendra BL, Wu H, Baccala R, et al. Discovery of peptoid ligands for anti-aquaporin 4 antibodies. *Chem Biol.* 2013;20(3):351–359.

24. Andrews PI, Dichter MA, Berkovic SF, Newton MR, McNamara JP. Plasmapheresis in Rasmussen's encephalitis. *Neurology.* 1996;46:242–246.

25. Adcock JE, Oxbury JM, Beeson D. Comparison of treatment of Rasmussen's encephalitis with plasmapheresis versus hemispherectomy. *Epilepsia.* 1997;30(S8):189.

26. Jaffer H, Adjei IM, Labhasetwar V. Optical imaging to map blood–brain barrier leakage. *Sci Rep.* 2013;3:3117–3124.

27. Wang S. Key to detecting Alzheimer's early could be in the eye. *The Wall Street Journal.* July 13, 2014.

28. Wu Z, Guo Z, Gearing M, Chen G. Tonic inhibition in dentate gyrus impairs long-term potentiation and memory in an Alzheimer's disease model. *Nat Commun.* 2014;5:4159.

29. Göbell B, Oltmanns KM, Chung M. Linking neuronal brain activity to the glucose metabolism. *Theor Biol Med Model.* 2013;10:50.

30. Liu-Ambrose T, Nagamatsu LS, Graf P, Beattie BL, Asche MC, Handy TC. Resistance training and executive functions. A 12-month randomized controlled trial. *JAMA Intern Med.* 2010;170(2):170–178.

Acetylcholine (Ach)   A transmitter of the cholinergic neurons secreted at the ends of nerve fibers to propagate a signal and was one of the first neurotransmitters to be identified. In the cortex of the brain, Ach plays a role in attention. Since the neurons that synthesize Ach die, less is present in AD, and therefore inhibiting the enzyme that breaks down Ach is a commonly used therapy to treat the symptoms of AD.

Alpha 7 nicotinic acetylcholine receptor   A ligand-gated calcium receptor stimulated by nicotine and acetylcholine; widely distributed in the nervous system; implicated in long-term memory.

Alzheimer's disease   A progressive, neurodegenerative disorder of the brain; coined by Dr. Alois Alzheimer based on the autopsy of a subject with severe memory problems, confusion, and difficulty understanding questions; neuropathological features include dense deposits of amyloid and neurofibrillary tangles in affected neurons.

Amyloid   Broadly described as insoluble aggregates of protein. Of the various types of amyloid, the β-amyloid ($A\beta$) form is associated with Alzheimer's disease that is a cleaved product from the amyloid precursor protein (APP). The $A\beta$ is further cleaved to produce the $A\beta42$ form that is implicated in the neuropathology of Alzheimer's disease chiefly because of its propensity to aggregate.

Apolipoprotein E (ApoE)   This protein combines with lipids in the body to form molecules called lipoproteins. These proteins package cholesterol and other fats and carry them through the bloodstream. Specifically, ApoE is a major component of the very-low-density lipoprotein type and removes excess cholesterol from the blood to transport to the liver for processing and therefore plays an important role in maintaining normal cholesterol levels.

Apoptosis   A regulated process of cell death that is in contrast to cells dying by necrosis, which happens from acute injury. In comparison to cell death by necrosis, apoptotic cells produce cell fragments called apoptotic bodies that phagocytes, like microglia, are able to engulf and quickly remove the contents of the cell before they cause further damage to the surrounding area.

Astrocytes   Star-shaped glial cells of the brain and spinal cord. They support the endothelial cells that aid in the maintenance of the blood–brain barrier. Other functions include a role in repair and scarring in the brain, as observed in the areas of the dense-core plaques, and are also responsible for the maintenance of extracellular ion balance.

Autoantibodies   An antibody that is directed against one or more of the individual's own proteins. They are created by the immune system when it fails to distinguish between self and foreign. Based on the data presented in this book, the AD blood–brain barrier that typically restricts access of antibodies to the brain fails to allow the unregulated entry of antibodies into the brain, some of which recognize and bind to specific antigens on neurons leading to their death.

Bapineuzumab   A humanized monoclonal antibody against the β-amyloid plaques to treat Alzheimer's disease but in 2012, failed to produce significant cognitive improvements in two major clinical trials.

Bielschowsky silver stain   About a 100-year-old staining method to use silver to nonselectively detect proteins in tissues and specific to the brain, it is used to demonstrate the presence of neurofibrillary tangles, nerve fibers, and senile plaques in Alzheimer's brain tissues.

Blood–brain barrier (BBB)   A physiological barrier comprising tightly joined endothelial cells to prevent certain substances in the bloodstream from entering the brain. Based on the "inside-out" hypothesis, the barrier is the first pathological event and root cause of Alzheimer's disease that leads to the unregulated entry of Aβ42 and neuronal-specific auto-antibodies into the brain.

Blood–retina barrier (BRB)   A physiological barrier comprising tightly joined endothelial cells to prevent certain substances in the bloodstream from entering the retina. The barrier becomes more porous in patients with diabetic retinopathy, which frequently occurs as the result of diabetes. Based on the "inside-out" hypothesis, the first cause of Alzheimer's disease is the breakdown on the integrity of the BBB that also hypothesizes that the initial pathological events in the BBB could be morphologically observed in the BRB through high-resolution eye exams as an additional AD risk factor while providing an assessment of the BBB without possible biopsies.

Cathepsin D   One of a family of cathepsin proteolytic enzymes that are located in the cell lysosomes and are chiefly involved in peptide synthesis and protein degradation.

Centromere DNA   A region of DNA typically found in the middle of a chromosome and represents large blocks of repetitive DNA.

Cerebral microbleeds (MBs)   Small chronic brain hemorrhages in the brain likely caused by abnormalities or damage of the small vessels.

Collagen IV   A type of collagen primarily found in the basal lamina of vessels and used in IHC to identify vessels.

DAPI   A blue fluorescent dye that binds strongly to the A-T rich sequences in DNA. DAPI is short for 4'-6'-diamidino-2-phenylindole.

Dementia   A general term to describe a decline in mental ability (e.g., thinking, memory, reasoning) severe enough to interfere with daily life.

ELISA   Abbreviated for enzyme-linked immunosorbent assay, which is a color test that uses specific antibodies to determine if a particular substance is present in a sample.

Entorhinal cortex   An area in the brain that functions as the major input and output structure of the hippocampus and also functions in memory and navigation. The entorhinal cortex is one of the first two areas of the brain vulnerable to damage at the early stages of Alzheimer's disease.

Epibatidine   The analgesic property of the poisonous alkaloid epibatidine is found in the skin of the *Epipedobates tricolor* species of the poison dart frog; is believed to pass through by binding to nicotinic receptors.

Formic acid   A carboxylic acid used to pretreat formalin-fixed brain tissues before the immunohistochemical methods to help the antibodies penetrate the antigen or target.

Galanin   A neuropeptide that is widely expressed in the brain and functions as a cellular messenger. In the book, galanin was used as a nonspecific peptide to test the staining properties of the Aβ42 antibodies.

Glial fibular acetic protein (GFAP)   A specific intermediate filament found in astrocytes, and used as a specific marker to identify astrocytes in immunohistochemical assays.

Gliosis   An inflammatory response by the glial cells (e.g., astrocytes, microglia) to damage in the brain. Initial response included the migration of the microglia to the site of the injury, followed by the production of a dense fibrous network of astrocytic processes producing the glial scar to isolate and sequester the damage from the unaffected areas in the brain.

Hematoxylin and eosin (H&E)   One of the principal and oldest stains in histology used to stain tissues and cells to aid in the morphological analysis of the sample. The gold standard is used routinely as the initial stain before more specialized staining methods were introduced. The hematoxylin stains all cell nuclei purple, and the eosin stains all other cellular structures (e.g., cytoplasm) pink.

Hippocampus   Part of the brain involved in memory forming, organization, and storing, spatial navigation, and important in forming new memories and associating them with emotions and senses. The hippocampus is one of the first two areas of the brain vulnerable to damage in the early stages of Alzheimer's disease.

Histology   The study of the microscopic anatomy of cells and tissues.

Histopathology   The microscopic examination of tissues and cells to study the manifestations of disease.

Human leukocyte antigen-D related (HLA-DR)   A cell surface receptor on inflammatory cells such as monocytes and macrophages, and specifically on the microglia in the brain. Antibodies specific to HLA-DR are used to detect reactive microglia in the brain.

Hypoperfusion   Decreased blood flow through an organ and if prolonged, may result in permanent cellular dysfunction and cell death.

Immunohistochemistry (IHC)   The process of detecting targets, or antigens, in cells or tissues through the use of specific antibodies and a detection system to color indirect presence of the target or antigen. IHC can detect the presence of one, two, or three simultaneous targets and is limited to number of unique colors available, and can be used in conjunction with other staining methods.

In situ hybridization   The process of detecting nucleic acid targets (e.g., DNA, RNA) in cells or tissues using labeled complementary probes to produce a color at its location.

"Inside-out" hypothesis   Proposes that the initial neuronal death in Alzheimer's disease is caused by unregulated intracellular accumulation of Aβ42 in neurons through the α7 receptor, leading to neuronal lysing and the development of dense-core plaques. This excess accumulation of Aβ42 in the brain is attributed to a dysfunctional blood–brain barrier, which should otherwise regulate and minimize the passage of amyloid. Once neuronal lysing and the formation of plaques begin, neuroinflammation and additional neuronal death follow.

Lipofuscin   An intracellular yellowish-brown, nondegradable pigment that remains from the breakdown of cellular components that remain in cells that do not divide. In the "inside-out" hypothesis, this indigestible material remains at the epicenter or middle of the dense-core plaque as the neuron dies.

Lysis   The breaking down, dissolution, or destruction of a cell by enzyme or other mechanisms to compromise the cell's integrity.

Lysosomes   Membrane-bound cell organelles containing enzymes (as many as 50 types), which are able to break down all kinds of biomolecules including proteins, lipids, carbohydrates, nucleic acid, and cellular debris. The membrane protects the rest of the cells from the degradative enzymes. In the "inside-out" hypothesis, it is the uncontrolled release of these active enzymes as the neurons die that created the spherically shaped, dense-core amyloid plaques.

Microglia   A type of glial cell in the brain that functions as the resident macrophage, or inflammatory cell.

Microtubule-associated protein 2 (MAP-2)   Intracellular protein associated with stabilized microtubules found in the neuronal dendrites.

Mild cognitive impairment (MCI)   Recognized as an intermediate stage of mild impairment between the expected cognitive decline in normal aging and the more serious decline of dementia. MCI may be perceived as a harbinger of Alzheimer's disease at the rate of 10–15% per year.

Necrosis   An irreversible form of cell injury that is caused by a number of factors and results in the unregulated digestion of its cellular components, much like what has been described for the origin of the dense-core plaques from neurons that overingest Aβ42. Necrosis can also be active in the inflammatory system.

Neurofibrillary tangles   Insoluble, intracellular aggregates of hyperphosphorylated tau protein found within some of the neurons in the AD brain.

Neuronal-specific nuclear protein (NeuN)   Expressed in most neuronal cell types throughout the nervous system, and often used as an immunohistochemical marker to identify neurons.

Neuropil (neuropile)   A general term to describe any area in the nervous system that includes the fibrous network of nerve fibers and glial cells between the neurons.

Parenchyma   Composed of the neurons embedded in the framework of glial cells (microglia, astrocytes, and oligodendrocytes) and blood vessels.

Perikaryon   The cell body area in the neuron that houses the nucleus and its organelles, and an area that begins to store Aβ42 over time.

Plaques   A general term to describe the morphology of amyloid deposition and neuronal fibers in the brain. Types of plaques include the dense-core plaques that are derived from bursting neurons, and diffuse plaques that are derived from amyloid that leaks into the brain from the vascular system.

Proteolysis   The enzymatic breakdown of proteins into smaller parts (amino acids, or polypeptides) through the hydrolysis of peptide bonds.

Pyknosis/pyknotic   The irreversible condensation or shrinking of the nuclear chromatin representing a degenerating cell.

Special stains   Typically refer to a group of staining methods, other than H&E, to help visualize and/or identify structures, cells, and cellular components in histological sections of tissues, or cell preparations.

Tau   Intracellular proteins that stabilize the cell microtubules. Mostly present in neurons of the central nervous system.

TUNEL   TUNEL is abbreviated for terminal deoxynucleotidyl transferase dUTP nick end labeling and is a staining method used to detect apoptotic cells based on their fragmented DNA.

Ubiquitin   An intracellular protein found in almost all eukaryotic cells (ubiquitously) that combines with obsolete proteins to make them susceptible to degradation. Also involved in the cellular processes of cell-cycle regulation, DNA repair, cell growth, and others.

Western analysis   Also known as Western blotting, Western analysis is a technique used to identify proteins in a sample lysate of digested and ground up cells or tissues, with the use of antibodies, and gel electrophoresis.

Edwards Brothers Malloy
Ann Arbor MI. USA
April 17, 2015